AFRIKA-STUDIEN Nr. 10

Die Schriftenreihe „Afrika-Studien" wird herausgegeben vom Ifo-Institut für Wirtschaftsforschung e. V. München in Verbindung mit

Prof. Dr. Heinrich Kraut, Dortmund
Prof. Dr. Dr. h. c. Rudolf Stucken, Erlangen
Prof. Dr. Hans Wilbrandt, Göttingen
Prof. Dr. Emil Woermann, Göttingen

Gesamtredaktion:

Dr. phil. Wilhelm Marquardt, München,
Afrika-Studienstelle im Ifo-Institut

Dr. agr. Hans Ruthenberg, Göttingen,
Studienzentrum für Tropische und Subtropische Landwirtschaft
in der Landwirtschaftlichen Fakultät der Universität Göttingen

IFO-INSTITUT FÜR WIRTSCHAFTSFORSCHUNG
AFRIKA-STUDIENSTELLE

African Agricultural Production Development Policy in Kenya 1952—1965

By

HANS RUTHENBERG

With 3 Maps and 35 Tables

SPRINGER-VERLAG
BERLIN · HEIDELBERG · NEW YORK
1966

SPONSORED BY THE FRITZ THYSSEN-STIFTUNG, KÖLN

Library of Congress Catalog Card Number 66-26098
Printed in Germany
Title-Nr. 7310

Summary of Recent and Forthcoming Publications under the African Research Programme

The whole research programme presently conducted by the African Studies Centre of the Ifo-Institute in connection with other institutes and research workers (see introductory remarks in no. 1 and no. 2 of the "Afrika-Studien") included the following investigations at the beginning of January 1966.

In order to keep readers informed of changes, supplementary work and the progress made towards publication each volume of "Afrika-Studien" will contain a summary of the programme as a whole.

General Economic Studies

a) Tropical Africa

N. Ahmad/E. Becher, Development Banks and Companies in Tropical Africa (printed as volume 1), in German

R. Güsten/H. Helmschrott, National Accounting Systems in Tropical Africa (printed as volume 3), in German

N. Ahmad/E. Becher/E. Harder, Economic Planning and Development Policy in Tropical Africa (completed), in German

b) East Africa

L. Schnittger, Taxation and Economic Development in East Africa (printed as volume 8), in German

R. Güsten, Problems of Economic Growth and Planning: The Sudan Example (volume 9), in English

P. v. Marlin, The Impact of Foreign Economic Relations for the Economic Development of East Africa (mimeographed in English)

R. Vente, Methods and Results of Economic Planning in East Africa (in preparation)

F. Goll, Israeli Aid to Developing Countries with Special Reference to East Africa (in preparation)

Agricultural Studies

a) Tropical Africa

(Joint Authors), The Effects of the European Common Market Regulation of Agricultural Products on the Export Trade of Developing Countries (mimeographed), in German

H. Pössinger, Results and Problems of Agricultural Development in Portoguese Africa (completed), in German

J. O. Müller, The Attitude of Various Tribes of the Republic of Togo especially the Ewe on the Plateau de Dayes to the Problem of Commissioned Cattle Herding by the Fulbe (Peulh) (being printed as volume 14), in German

b) East Africa

1. Country Reports

H. Ruthenberg, Agricultural Development in Tanganyjka (printed as volume 2), in English

H. Ruthenberg, African Agricultural Production Development Policy in Kenya 1952—65 (volume 10), in English

2. Studies Concerning Grassland Use and the Cattle Economy of East Africa

H. Leippert, Natural Planting Associations in the Semi-arid Areas of East Africa (in preparation)

K. Meyn, Beef Production in the Semi-arid Areas of East Africa (in preparation)

E. Raddatz, Smallholder Land Use and Milk Production in Kenya (in preparation)

H. Klemm, The Organization of Milk Markets in East Africa (in preparation)

3. Studies in the Organisation of Smallholder Farming in East Africa

D. v. Rotenhan, Wasukuma Agriculture, the Organisation of Cotton Farms in Sukumaland, Tanzania (being printed as volume 11), in German

H. Pössinger, The Possibilities and Limitations of Smallholding Sisal in East Africa (being printed as volume 13), in German (mimeographed in English)

S. Groeneveld, Traditional Farming and Coconut-Cattle-Schemes in the Tanga Region (in preparation)

W. Scheffler/A. v. Gagern, Smallholder Tobacco in Tanzania (in preparation)

K. Friedrich, Bahaya Agriculture, the Organisation of Coffee-Banana-Farms in the West Lake Region, Tanzania (in preparation)

E. Baum, Traditional Farming and Land Development in Kilombero Valley, Tanzania (in preparation)

4. Other Research Concerning Agricultural Development

M. Paulus, The Role of Co-operatives in the Economic Development of East Africa, and especially of Tanzania (completed), in German

N. Newiger, Co-operative Farming in Kenya and Tanzania (completed), in English

M. Attems / W. Poeplau / Chr. Schlage, Relationships between Health, Nutrition and Agricultural Development. Three case studies from the Usambara-Mountains/Tanzania (in preparation)

F. Dieterlen/P. Kunkel, Tropical Rodents and Birds as Agricultural Parasites completed, in German

W. Kühme, A Study of Animal Behaviour in the Serengeti completed, in German

Studies in Commerce, Trade and Transport

H. Kainzbauer, Trade in the Economic Development of Tanzania (in preparation)

W. Lamade, The Development and Role of Marketing Boards in Tanzania (in preparation)

K. Schädler, Handicrafts in the Economic Development of Tanzania (in preparation)

H. Reichelt, The Structure and Level of Costs in Kenya's Industry (in preparation)

G. Rötzer, Transport Problems in the Forest and Timber Industry in East Africa (in preparation)

Sociological Studies

A. Molnos, Methods and Results of Sociological Research in East Africa 1954—1963 (printed as volume 5), in German

A. Molnos, The Status and Role of Women in the Changing Economic and Social Structure of East African Rural and Urban Areas (in preparation)

O. Raum, The Human Factor in the Development of the Kilombero Valley (mimeographed), in English

O. Neuloh a. o., The African as a Worker in Industry in East Africa (in preparation)

H. Jürgens, Contributions to Domestic Migration and Population Development in Liberia (printed as volume 4), in German

I. Rothermund, The Political and Economic Role of the Asian Minority in East Africa (printed as volume 6), in German

E. Ortenau, The Introduction of Economic Innovations among the East African Tribes (in preparation)

E. Gerken, The Sociological Change in Rural Communities of Uganda under the Influence of Industrialisation and Urbanisation (in preparation)

Legal Studies

H. Fliedner, Land Tenure Reform in Kenya (printed as volume 7), in German

H. Krauss, Land Legislation in the Cameroons 1884—1964 (being printed as volume 12), in German

Chr. Fuchs-Stubenrauch, The Present Position of Legislation in East Africa (in preparation)

Studies in the Field of Economic Geography

W. Marquardt, The Interrelationships between Man, Nature and Economic Organization with special reference to Madagascar (in preparation)

H. D. Ludwig, Ukara — a Development Study in Economic Geography (in preparation)

R. Jätzold/E. Baum, Studies in Economic Geography in the Kilombero Valley/Tanzania (in preparation)

Bibliographies

D. Mezger/E. Littich, Recent English Economic Research in East Africa. A Selected Bibliography (in preparation)

A. Molnos, op. cit. Sociological Studies

Preface

In volume 2 of the "Afrika-Studien" an appraisal of the agricultural development policy in Tanganyika (now Tanzania) from 1950 to 1963 was made. This report is a continuation of the work in East Africa, with the emphasis lying on a tentative quantitative assessment of costs and benefits of smallholder development. There are few countries in Africa south of the Sahara where as many and as various measures for the promotion of smallholder farming have been tried as consistently and intensively as in Kenya. In particular the "Swynnerton Plan" led to the employment of substantial sums in African farming. Some of the approaches have been highly successful, others not. It is the purpose of this report to inform about aims and institutions, methods and difficulties, costs and benefits.

Prof. Dr. EMIL WOERMANN
Institut für landwirtschaftliche
Betriebs- und Landarbeitslehre,
Göttingen

Acknowledgements

As usual with reports of this nature, my main debt is to a great number of smallholders, settlers, scheme managers, Agricultural Officers and Instructors who so willingly discussed their problems with me and thus provided the information on which this report is based. I am most grateful for the support rendered by the various Departments of the Kenya Government. A debt of gratitude is owed particularly to the Fritz Thyssen Foundation, Cologne, which provided the funds and to the Ifo-Institute, Munich, which provides the institutional framework for German economic research work in East Africa. Most valuable advice and criticism was given by Mr. J. D. MACARTHUR, Mr. L. H. BROWN and Mr. D. M. ETHERINGTON. Thanks are also due to Mrs. M. P. ETHERINGTON who corrected my English draft.

HANS RUTHENBERG

Contents

Tables

Maps

A. Introduction

Kenya is one of the few undeveloped countries where agricultural production has been rising significantly within the last decade. The total agricultural production increased at an annual rate of 4.5 per cent, from £ 74 m. in 1954 to £ 117 m. in 1964. It should be noted that both sections of the agricultural economy, the large farms as well as the smallholders, contributed almost equally to this increase. The standards of farming in Kenya have reached a level far beyond those achieved in the other East African countries. The large farms and plantations are among the best in the tropics, and the quality of farming in some smallholder areas of Kenya is certainly unchallenged anywhere else in Africa south of the Sahara.

Most of this has been achieved through the provision of public funds for the development of individual holdings. The expenditure of the Ministry of Agriculture, which accounts for almost all the money that went into the promotion of smallholder farming, and for a significant part of the services and finance provided for large farms, oscillated in the relevant years between £ 4 m. and £ 7 m., that is to say from six to nine per cent of the value of total agricultural production. If we compare the Ministry's additional expenditure between 1952/53 and 1960/61 and relate it to the additional production from 1956 to 1964 (thus assuming a time lag of three and a half years between expenditure for agricultural services and production response), we obtain a relationship of 1 : 8.4. This means that the additional expenditure of £ 1 for agricultural development, under the prevailing circumstances, resulted in additional agricultural production of £ 8.4.

The above figures indicate not only that agricultural development in Kenya achieved notable rates of growth, but also that it has been expensive, in particular if we consider that the large farm economy financed most of the expansion out of reinvested profits or imported capital.

Public funds were spent on a wide range of measures for the promotion of African farming. It is important for agricultural development in Kenya and elsewhere to have an approximate idea of their costs and benefits. This report is an attempt to evaluate the experience made so far.

Table 1. *Small farm production within total agricultural production*

	Expenditure[1] for Agr. & Related Services[2,4]	Total Prod'n	Total Small Farm Prod'n	Small Farm		Small Farm Production	
				Marketed Prod'n[3]	Subsistence Prod'n	As % of Total Production	As % of Marketed Production
	£m	£m	£m	£m	£m	%	%
1952	4.1	..[8]	..	..	..	..	..
1953	4.0	..	..	..	..	..	..
1954	6.0	74.0	52.2	6.6	45.6	70	24
1955	6.6	74.9	52.0	6.2	45.8	69	22
1956	5.6	83.0	54.1	6.1	46.0	65	18
1957	4.8	85.3	58.6	6.9	51.7	69	17
1958	5.8	87.2	60.4	7.8	52.6	69	19
1959	5.7	88.6	61.4	8.4	53.0	69	20
1960	6.8	90.2	61.8	9.5	52.3	69	20
1961	7.8	86.6	60.0	10.4	49.6	69	23
1962[5]	11.4	103.1	75.4	10.5	64.9	73	22
1963	15.5	111.0	79.0	11.6	67.4	71	22
1964[6,7]	..	116.9	83.4	14.0	69.4	71	25

[1] The fiscal year is from July to June.

[2] Agriculture includes incidental services, livestock, forestry, fishing, and hunting.

[3] Marketed production of small farms is defined as that part of production which is sold outside the growing district. On the farm level, the proportion of marketed production is higher.

[4] One third of the actual "Emergency" expenditure for soil conservation work was brought into the total, because it can only partly be considered an economic service.

[5] The jump in subsistence production between 1961 and 1962 is partly due to a different method in estimating smallfarm production.

[6] Provisional.

[7] Small farm data include production in resettlement schemes which constitutes 2 per cent of agriculture's contribution to the Gross Domestic Product.

[8] In this and other tables the following symbols have been used: "—" means "nil"; "..." means "figures not available".

Source: Kenya: Economic Survey 1965, Nairobi 1965. See also the Economic Surveys of the preceding year. — Kenya: Statistical Abstracts, Nairobi 1964. Table 123 (and preceding years).

B. Stages in Development Policies

I. The Establishment of the White Highlands

The first step in the agricultural development of Kenya was the planned establishment of a dual society. The land of Kenya, comprising 220,000 square miles altogether, but only 38,000 square miles suited for arable farming and forestry and 20,000 square miles with a marginal agricultural potential, was split into two segments. The greater part, comprising not only extensive low-potential areas of bush and semi-desert, but also the greater part of the high-potential land, — the estimates vary between two-thirds and more than four-fifths — and definitely almost all the best land remained available for the African peasants and herdsmen. These African lands were called "Non-Scheduled Areas", or the "Reserves". The smaller part, the so-called "Scheduled Areas", or "White Highlands" [1], amounting to 7.6 million acres, was made available to 3,600 European farmers (some areas on the coast and in Nyanza have also been open to Asian farmers) who, in due course, helped by loans and ample labour, established a belt of modern farming stretching from Nairobi to the Uganda border [2].

There were several reasons for the establishment of such a foreign enclave:

- The railway, built largely for non-economic reasons, had to be made economic by export production. European large farms were an obvious way of achieving this objective.
- There was empty land in a healthy climate, which seemed to be no man's possession. No titles or permanent habitation existed, and the only tribe that could have laid claim to much of it was the Masai who lost great numbers of men and cattle at the turn of the century, due to drought and disease, and who could not make use of it.
- The Kenya Government agencies wanted British settlers to come and there were enough people, who liked the outdoor life which farming in Kenya offered. There is historical evidence that the settler community had sufficient political influence to muster support from the government back at home.
- There was the chance of making colonial administration easier by placing European farms as buffer zones between feuding tribes.
- There were — and still are — compelling reasons to group large farms together: control of plant disease, animal disease and theft; easier access to markets, creameries and social centres; easier exchange of ideas and innovations, etc.

---

[1] Not all of the "Scheduled Areas" were in the "White Highlands". Some stretches of land on the coast and in Nyanza (Asian sugar estates) were also alienated for large-scale farming.

[2] For a detailed discussion of the agricultural potential, the types of land-use in Kenya, the traditional land tenure and the land question, see the bibliography and in particular the East Africa Royal Commission Report, 1953—1955.

Agricultural development policy, that is to say, provision of loans, administrative services, roads, market outlets, and research, was, up to the mid-thirties almost exclusively, and up to the mid-fifties largely, a matter of promoting and firmly establishing the White Highlands. The idea,

"High Rainfall Areas in Kenya"

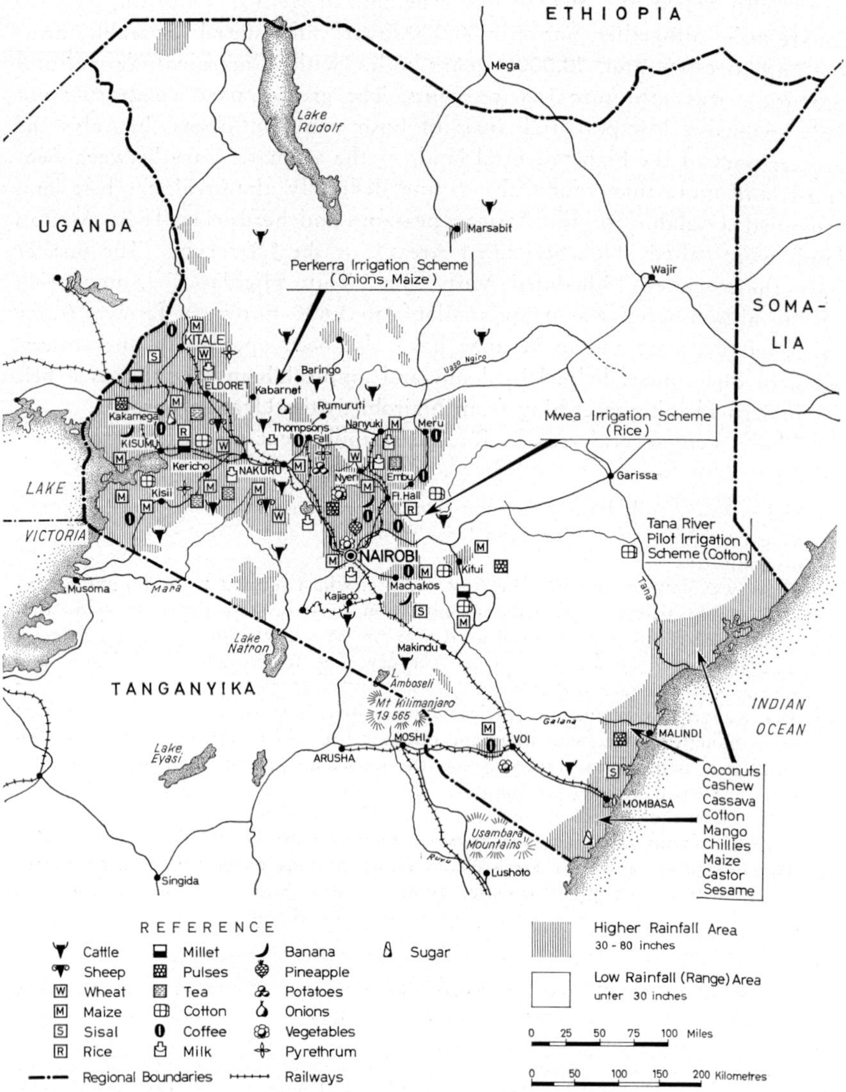

IFO-INSTITUT für Wirtschaftsforschung München 54/66

certainly a sound one, behind this policy was that Government agencies were no effective substitute for the skill of a large number of competent agricultural entrepreneurs. The foreign enclave was, in due time, expected to benefit not only the European farmers but also the Africans, by giving employment, spreading skills and innovations, opening markets, and creating tax revenues with which to finance social services. There is no doubt that the establishment of a dual farming society was highly successful economically. The spread of innovations and improvements, the rapid application of worthwhile results of research, was something that the European farmers largely took care of for themselves. The conditions for development were favourable indeed: a small group of agricultural entrepreneurs with a high educational standard, a common language, a rather homogeneous outlook on economic affairs, and a closely-tied system of social connections. The large farms became the economic backbone of the country, contributing, between 1954 and 1963, one-third of the total agricultural production and four-fifths of market sales. Their economic activity largely supported the remainder of the economy and most of the social services for the Africans, such as schools, hospitals, roads, etc. Thanks partly to research, but mainly to entrepreneurial activities, most of the technical knowledge needed for farming anywhere in Kenya is now readily available. Major roads and market outlets have been established.

It can hardly be denied that the existence of the "White Belt" within Kenya smoothed the path for African agricultural development to a degree which is probably unique. At the same time, it created a host of severe social problems. The British administration tried to adhere to the principle laid down by the Colonial Office: to put African interests first. There was certainly no deliberate neglect of the Non-Scheduled areas. There was, however, a difference in emphasis. In the European areas, owing to the set of values to which the large farmers and businessmen adhered, efforts were focussed on building up profitable enterprises with little regard for the cultural problems of the indigenous population. The African areas, on the other hand, were the responsibility of well-meaning administrators with a less dynamic set of values who, as elsewhere in the British colonies, saw the African interests — at least up to 1950 — as being the conservation of natural resources and the preservation of a traditional way of life rather than social change. This is certainly not the fault of the administrators who in this way merely supported the wishes of the people, but it must largely be attributed to the socio-political setting of the time.

There was the need for revenues and railways traffic, and as early as 1900 some efforts were made to promote market production, for instance, in cotton and maize in Nyanza Province. In Meru, Embu and Kisii some pressure had to be exerted in the 1930's to get coffee planted. But up to the mid-thirties or even the mid-forties, no systematic policy could be devised for the development of the African smallholdings. The Agricultural Officers of that time did most valuable work in guiding progressive farmers as much as possible and in accumulating information about the technical, economic and social conditions of farming in the Reserves. In many respects they laid the foundation for the successful approach in the late fifties. But there were few of them. Funds were scarce and the direction of their work was definitely non-commercial. Furthermore, any systematic policy for developing the potential of the African areas would have been bitterly fought by the African elders and politicians themselves. No one could even contemplate land consolidation in those days. A peasant who wanted to develop his own farm was — as a rule — opposed bitterly by his own people. The rural population who had the choice between working on a large farm or living on one's own holding usually thought the former more rewarding, and many smallholders left the farm under the care of their wives, who continued to grow subsistence crops as laid down by age-old traditions. Consequently, the Non-Scheduled areas, though containing almost all the best land of Kenya, developed into reserves supplying cheap and ample labour for the large farms, and part of the maize, and the low-grade cattle with which the farm labour force was fed.

II. African Land Development up to the Early Fifties

The Labour Government in Britain with its Ten Year Plan for Kenya, starting in 1946, brought about a significant change. The large farm sector continued to be considered the mainstay of the economy and received most of the attention. However, more money and personnel were made available for African areas. The policies of that time were still largely influenced by the non-commercial attitudes of the Ministry of African Affairs and their conservative counterparts, the elders in the Reserves. Not the more technical-minded officers in the technical departments, but the Chief Native Commissioners, exerted the greatest influence. They thought — following the thinking of their time — that it would not be in the interest of the local people if they were to become rapidly integrated in the workings and the attitudes of the commercial world. The policies pursued in the early fifties followed the analysis of the problems of African farming:

— There was ample evidence to consider soil erosion to be the greatest danger. The 1951 report on Agricultural Policy states: "Our first aim must be to prevent further deterioration of the land". "Paramount importance" was given to the "immediate and vigorous execution of projects for the conservation of the soil and the preservation of water supplies." Persuasion was tried, but failed, partly because there was not enough staff. Therefore, the African District Councils were made the agents of the Government's policy. They had to pass by-laws making, for instance, the terracing of steep land obligatory. The local authorities obtained the legal power to call on all able-bodied men to perform services for the community six days per quarter and to compel the individual to work on land improvement schemes. Consequently, tens of thousands of miles of terraces were built on the overcrowded hillsides under government pressure or even compulsion, with the policeman standing behind the Agricultural Instructor.

— The only answer to any apparent overcrowding in some African areas was seen, apart from soil conservation practices, in the removal of people and stock from denuded and eroded areas. A settlement agency was established. Substantial sums were invested in the opening up of lands which should absorb surplus population from the hills in an orderly manner.

— Some attention was given to the introduction of cash crops. The first nurseries for coffee, tea and pyrethrum were established. The main emphasis, however, was laid on better husbandry of subsistence crops. The 1951 Report of the Department of Agriculture states: "People like to forget about food and to concentrate on a quick cash return" [1].

— Particular emphasis was given to group farms and group activities, partly because individual holdings seemed too radical a change to some agricultural officers, and partly because co-operative or communal efforts fell in nicely with the thinking of the British Labour Government at that time.

Useful as these efforts might have been, they did little to stimulate any activity within the peasant community. Group farming failed because the people concerned mistrusted each other and thoroughly disliked the idea. The lazy man benefited more than the hard worker, who soon gave up. The priority attached to improving subsistence crops, which prevented a great deal of hunger, failed to arouse the farmer's interest. In any case, most of the work for food production was done by the women, who hardly ever came in contact with the extension worker. The introduction of ley systems proved too difficult and unrewarding a step to begin with [2]. The settlement absorbed a lot of money without noticeably relieving the land pressure. Soil conservation work, which undoubtedly healed much of the erosion damage and has definitely changed for the better the face of several parts of the countryside, — parts of Teita and Machakos District for instance — met with opposition from the people. There was lack of insight, dislike of

[1] The relative emphasis attached to either subsistence food production or cash (export) production changes according to circumstances. Up to the 1950's the emphasis was on subsistence. During the 1950's cash crops stood in the foreground. In the mid-1960's food shortages developed, partly due to the allocation of labour and land to cash crops, and efforts concentrate again on feeding the people.

[2] The idea was to subsistute planted grass for the traditional tumbledown fallow so that a milk cow could graze and provide protein for the family. There are many adequately good local grasses, and the planting of grass fits in nicely into the African farming systems, which are based on a rotation of fallow and arable use of the land. It is, however, a matter of discussion whether a ley system indicates the expansion path of smallholder farming in densely populated areas. The developments which take place on Mt. Kilimanjaro and in some parts of Uganda indicate that zero-grazing might be more appropriate.

communal work, and in particular, hatred for the foreign overlords who compelled the people to work for no apparent benefit to the Africans themselves. It is likely that the pressure to carry out soil conservation work was one of the factors which touched off the Emergency[1].

A case in point were the conditions in Nyeri District, Central Province. The Annual Reports of the agricultural district office 1952 and 1953 comment: "They (the farmers) not only do not want to progress themselves but want to stop everyone else progressing ... Everything sponsored by the European is wrong and has to be abandoned ... If the Kikuyu is to progress he must be spoon-fed ..." Opposition was particularly strong against soil conservation work. FISHER reported: "... all people feel bad because Europeans tell them to dig contour ditches ... the old did not think the soil was spoilt because they harvested a lot of crops of all kinds" (p. 282). The Annual Report of 1952 adds to this: "... people are loath to give up land for drainage ways ... there have been numerous incidents of political agitation directed against soil conservation work ... people decide to ignore by-laws passed by the District Councils ...".

III. The Emergency and its Economic Consequences

The Mau Mau Emergency from 1952 to 1955 was caused by many factors, predominant among which were:

- the fear that European domination, established by settlers who wanted to make permanent homes in Kenya, might be something that would last for an indefinite time;
- the conviction that European wealth was built on land taken away from the African and on cheap African labour;
- the over-population in some districts in relation to the traditional system of land use;
- the animosity towards the break-down of tribal customs; and
- the pressure which the Government was exerting on the rural population in its efforts to get soil conservation work done.

The Emergency had decisive and — from a purely economic standpoint — beneficial consequences. Political stability can be considered to be the very prerequisite of economic development, and when the Emergency ended a long time of assured British rule seemed to lie ahead. Investors were confident, and investments on large farms reached an all-time high figure of £ 5.9 m. in 1960. A substantial part of the modern sector of the economy originates from the fact that the Emergency focussed attention on

[1] Soil conservation was regularly and sometimes violently opposed by politicians. The people could often see the sense of it, but simply did not like the extra work and could easily be persuaded by the politicians to down tools. However, the dangers of erosion are evident everywhere. The farmers do not like the extra work. Consequently, it is most likely that the present Government — sooner or later — will be compelled to exert pressure in order to prevent starvation, i.e. to repeat the process which had to be given up by the British.

Kenya, that aid measures furthered internal buying power, that additional Government loans were provided, and that the suppression of Mau Mau was considered to have strengthened the British position.

The Emergency, furthermore, brought drastic changes in the attitude towards African farming. The old agricultural policy which ran under the headline "development of the large farm economy first" and "protection of the African interest in the Reserves within the traditional way of life" came to an end. The years 1955 to 1962 were characterized by rapid progress in both farming communities, the European as well as the African. The Emergency gave support to those who for some time already had advocated an active policy of smallholder promotion. The way to develop the economy more rapidly and at the same time to secure properties in the White Highlands was seen not only in suppressing anti-European activities, but also in making the African realise the income potential of the land in the Reserves.

The increase in Government expenditure coincided with a favourable psychological situation. The effects of the Emergency can, in some respects, be compared to those of a lost war. The defeat of the political aspirations of many of the Kikuyu, Wameru and Embu left only one outlet for unsatisfied but active people: agricultural production. There was an increasing willingness to act according to the demands of a modern economy, a factor which is often observed in the wake of a major social upheaval. In addition, the spread of education, the construction of more and better roads, and the increased buying power of the internal market — in particular the buying power of the labour force employed on large farms and plantations — brought more and more smallholders into the exchange economy.

The Emergency, furthermore, made it easier for the British to enforce those measures thought to be appropriate. The need for drastic changes was evident. One no longer feared to push aside traditional customs. The British were — at least in Central Province — in a position of absolute authority and thus able to carry out and control development measures with hardly anybody in a position to object.

The new approach, formulated in the Swynnerton Plan and backed by the substantial sum of £ 5 million, put into effect on a large scale what pioneering Agricultural Officers had advocated or shown long before to be workable solutions. The new approach was characterized by

— a *land reform* bringing land consolidation and the issuing of title-deeds, in particular to Central Province where it was largely welcomed by the people and aggressively pushed by the Government. Land consolidation in Central Province took six years instead of the planned fifteen to twenty years. Even at the height of the Emergency, voluntary demand for land consolidation in Central Province was higher than the available staff could deal with, and those who disliked the change in land tenure had no chance of voicing their opposition under the conditions prevailing at that time. The impact on the farmer is summarized in the 1958 Annual Report of Nyeri District: "The African landowner is prepared to work hard to improve his property, to make it productive and so to raise his status amongst his fellow farmers. He was not prepared to do this before land consolidation, as there was always the fear that he might lose his land after improvements had been carried out, through bribery or jealous relations who were not prepared to see any advancements in their midst." In several other parts of Kenya land reform took the form of voluntary enclosures, with Government seeking to couple it with adequate measures to ensure soil conservation, access roads, and water supplies for livestock. In some districts, registration followed and still continues to follow on the heel of voluntary enclosures. In March 1963, 1.4 million acres were already consolidated and 2.5 million acres enclosed — almost all of it land of high potential.

— changing *the emphasis from the development of subsistence crops to the introduction of new cash-yielding enterprises.* In the past it had been widely held that the African farmer was not yet capable of producing plantation crops on the same scale, of the same quality and with as sufficient a degree of disease control as the European. In some districts, restrictions were such that it was

Table 2. *Small farm area — value of gross farm revenue derived from marketed production* [1] *(1954—1964)*

	£'000										
	1954	1955	1956	1957	1958	1959	1960	1961	1962	1963	1964 [2]
Coffee	172	280	322	894	1,266	1,830	2,247	2,816	3,295	3,022	5,366
Pyrethrum	40	86	77	61	89	200	484	611	488	359	492
Tea	—	—	—	—	—	—	30	26	54	122	159
Sisal	15	6	23	5	3	65	284	435	132	519	667
Rice	—	—	—	—	138	103	265	384	314	320	343
Tobacco	—	—	—	47	52	35	39	58	66	111	76
Wattle	—	—	—	307	267	193	231	147	220	268	236
Maize	—	—	—	1,180	1,118	1,006	960	1,082	1,131	1,244	749
Cotton	591	487	518	417	348	499	545	505	271	498	550
Oil Seeds	—	—	—	—	329	173	241	216	284	337	200
Pulses	147	73	93	225	373	294	307	312	251	282	321
Coconuts & Products	—	—	—	56	236	405	289	286	215	246	324
Potatoes	70	107	93	107	60	96	125	199	105	132	151
Cashew Nuts	30	34	51	85	57	120	193	252	68	113	221
Fruit	—	—	—	—	193	188	175	103	228	230	173
Vegetables	—	—	—	—	—	—	259	308	167	203	231
Cattle and Calves [3]	—	—	—	1,800	1,876	1,848	1,870	1,760	2,120	2,184	2,258
Sheep and Goats	—	—	—	218	242	248	253	252	306	316	319
Poultry and Eggs	—	—	—	51	64	74	57	62	82	54	85
Dairy Products	—	—	—	199	200	178	193	246	286	518	691
Pigs [3]	—	—	—	185	153	25	8	5	9	5	21

[1] Marketed production is defined as that portion of production which is sold outside the growing district.

[2] Provisional.

[3] For slaughter.

Source: Kenya: Economic Survey, 1965 Nairobi, 1965, Table 23 — See also preceding years.

practically impossible for smallholders to plant coffee or tea [1]. The Swynnerton Plan brought a sweeping change. The future policy was based on the idea that smallholders can be efficient producers of plantation crops, provided that the growing is closely supervised by competent officers with adequate power to ensure that cultivation and all other aspects of production are carried out properly. The direct approach to soil conservation was replaced by an indirect one. Instead of compulsory terracing, we find close supervision of good husbandry — including soil conservation — of cash crops and grade cattle. The idea of group farming, which had not met with any response from the people, was replaced by approaching and aiding the interested individual farmer and his holding.

— efforts to *reorganise land use in pastoral areas* by establishing grazing schemes which brought with them (1) the adjustment of stock carrying capacity, (2) veterinary supervision, (3) controlled grazing, and (4) the organisation of market outlets.
— some *settlements on rain-fed land* and some investments in irrigation farming and settlement. The relative importance attached to settlement was, however, far less than it had been in the earlier phase.
— the *provision of more finance and personnel* for *auxiliary services* and technical assistance: nurseries, water supplies, education, transport, marketing etc.

Table 3. *Per cent of total marketed production derived from the small farm sector* [1, 2] *(1954—1964)*

	1957	1958	1959	1960	1961	1962	1963	1964
Coffee	9	13	19	23	27	29	30	36
Pyrethrum	6	8	14	20	25	24	23	50
Tea	—	—	—	1	1	1	2	2
Sisal	—	—	2	6	10	3	7	9
Wattle	47	37	33	48	29	33	46	43
Maize	32	32	39	35	39	33	43	42
Fruit	—	32	24	21	15	30	32	28
Cattle	49	47	44	42	39	38	40	47
Sheep and Goats	62	65	58	56	50	49	54	57
Poultry	15	16	20	14	14	16	13	23
Dairy Products Milk Etc.	4	5	4	4	5	6	10	14

[1] Total Marketed Production is that of small farms plus Gross Revenue of large farms.

[2] Marketed production of small farms is defined as that part of production which is sold outside the growing district.

Source: Kenya Economic Survey 1965, Nairobi 1965 p. 24 ff. See also preceding years.

---

[1] In every district there were a few Africans who grew plantation crops of high quality and who were used by the Agricultural Officers to base policies on possible performance. The argument before the Swynnerton Plan was against a rapid spread of cash crop production without the personnel to supervise the plantings. That the argument had some degree of validity is borne out in recent years, when control has been relaxed. The state of African coffee has become poorer, and the disease situation in some areas is worse than on the plantations.

The smallholder development policies pursued after the Emergency resulted in an impressive degree of response from the rural population. Total smallholder production, and in particular market production, increased rapidly (see Tables 2, 3 and 4). Most spectacular was the rise in coffee production from £ 172,000 in 1954 to more than £ 5,000,000 in 1964. Significant progress was made in the production of pyrethrum, tea, sisal, rice, milk, and several other crops. Within an amazingly short period, smallholder market production became a major industry in Kenya.

IV. Independence and the Africanisation of Kenya's Agriculture

The most recent phase of production development policies, beginning in 1962 — shortly before independence — is characterized by a continuation of the policy introduced after the Emergency with the Swynnerton Plan but now carried out with more and more Africanised services, and the Africanisation of the former White Highlands, thus bringing to an end the racial split in landownership which prevailed for half a century in Kenya. The Africanisation of the large farm economy is taking mainly three forms.

— In some areas, in particular those bordering on thickly populated former Reserves, large farms are bought and split up into numerous smallholdings which are given to African settlers. This approach is largely confined to mixed farming areas.

— Individual Africans, usually sponsored by family groups, and cooperatives are buying large holdings, supported by loans from the Land Bank and other sources.

— In areas which are to be opened up for arable farming or ranching, the emphasis is no longer on private European or Asian entrepreneurs who want to establish permanent homes in Kenya, but on state enterprise by the newly established Agricultural Development Corporation or on joint ventures between the above body and private firms.

In mid-1965, when this report was written, the Africanisation of Kenya's agriculture was in full swing.

C. A Benefit-Cost Analysis of Different Approaches to Agricultural Development in Kenya 1952—1965

It is generally assumed that the provision of agricultural services, aiming at the economic stimulation of existing peasant communities, is the most promising approach to agricultural development, not only in social or political terms but also from the point of view of the Treasury which has to make the funds available and to sustain further development efforts out of the returns from these investments. The experience in Kenya is considered to be a case in point. There is a demand for more agricultural services, posing the question as to whether it is more economic to invest, for instance, in extension and related services or in specific schemes: tea-development, irrigation, settlement on new land, re-settlement etc. As a contribution to the discussion taking place, the following evaluations and estimates are made of the costs and benefits of the more important approaches to African agricultural development in Kenya from 1952 to 1965. The estimates are necessarily rough because information about production volume and values and about expenditure is fragmentary.

I. Agricultural Administration and Extension

1. Aims, Institutions and Methods

a) Aims

In its last decade British colonial policy in Kenya aimed primarily at bringing about a rapid rate of indigenous economic development. The emphasis was on smallholder agriculture; the main instruments, the agricultural and veterinary administration, and extension services. As regards the direction of work we can distinguish between technical, economic, social, and political objectives. The principal technical objectives were:

- a cash enterprise for each farm;
- introduction of innovations: better seeds, better methods, better tools, fertilisers, insecticides, etc.;
- proper husbandry, in particular of cash crops and grade cattle;
- soil conservation;
- development of a sound system of farming with the emphasis on the establishment of leys and the integration of animal husbandry and arable farming.

Widely held, although not explicitly stated or planned, *economic objectives* were:

— provision of subsistence and a cash income of £ 100 annually for each farm;
— development of the potential productivity of the Non-Scheduled areas in order to obtain more marketed production and all the benefits which would accrue from this;
— expansion of the rural market for non-agricultural products in order to widen the scope for local industries, and incidentally, to increase excise and customs revenue;
— an increase of total revenues for the African District Councils, because of increasing pressure of the people for more schools, hospitals, feeder roads, and water supplies. The search for finance induced the Colonial Government to make additional efforts to realize the potential for agricultural production existing in the Reserves.

In addition, there were *social objectives*, for instance:

— the expectation that smallholder development would productively absorb surplus population and thus reduce the pressure of unemployment in towns;
— the hope that there would result a speed-up in social change. In particular, the cash incentive connected with cash crop production was considered a powerful means of adapting the way of life to the demands of a modern exchange economy.

Finally, there were important political objectives:

— A prosperous middle-class society of farmers, firmly established on the land, was expected to exercise a stabilising influence on politics in Kenya.
— It was hoped that the development of the vast potential of the Non-Scheduled areas would lessen the desire for land in the Scheduled areas and would thus give long-term stability to the large farm economy.

As important as the economic, social and political objectives might have been, they were kept largely in the background. The staff of the Department of Agriculture were mainly concerned with the realisation of the technical objectives, in the belief that a rapid rate of technical progress must necessarily improve the economic, social and political situation. This has been well described by D. J. Penwill:

"... The whole of this ... represents our local attempt ... to accomplish a social revolution; to try to help the transition from The Iron Age ... to standards of life and social behaviour set by the immigrant European community ... The field officer's task is to improve the husbandry standards in every way, guide the communal effort, organise nurseries for new cash crops, and generally to implement the betterment schemes, in detail, on the ground ... These field officers have no paper work, save to write a longhand report and pay out staff ... which should not take up more than two days. The rest of the time they spend out in their locations ..."

b) Institutions

Agricultural development within a smallholder community depends mainly on a more efficient use of existing resources. Land, water, labour, and even some capital, in the form of cattle, trees and hoarded savings, are usually available. The main stumbling blocks are the political, social and technical difficulties of introducing improvements and innovations which make agricultural work more remunerative. Smallholder communities tend to continue traditional patterns of farming in spite of the fact that the population and its needs are increasing rapidly. This leads, in due course, to the reduction of fallow, the cultivation of marginal land and steep slopes, low yields per acre, poor cattle, and increasing erosion. African peasants, as opposed to entrepreneurs with large farms, are hardly in a position to help themselves because they (1) are bound by tradition, (2) lack the cultural heritage which is behind proper husbandry, (3) depend on their neighbours, (4) shy away from risks, in particular where they would have to introduce a whole sequence of innovations at once, (5) do not have the initial finance, and (6) have little access to auxiliary services such as nurseries, processing facilities, markets, credit, etc. They need information, guidance and some degree of supervision, which in Japan, for instance, was given by the village landlords, and which, in former British Territories, was and still is more or less provided by the field staff of the Department of Agriculture and of other government departments providing agricultural services. The work of the field staff is usually called extension work, although most of the time is taken up by purely administrative duties, and consists of technical services, outright supervision, legal enforcement of land ordinances, etc.

In Kenya each Province is divided into administrative Districts with an Agricultural Officer (A.O.) in charge, who is usually a college graduate. Under the Agricultural Officer are Assistant Agricultural Officers (A.A.O.s.) who work in the various Divisions of the Districts. These officers have usually received two years of agricultural education. Within the Divisions are Agricultural Instructors (A.Is.) and Assistant Agricultural Instructors (A.A.Is.), who are responsible for extension and supervision at farm level. In 1962, the agricultural field staff in Kenya numbered 179 A.Os and A.A.Os., 1,107 A.Is. and 2,347 A.A.Is., of whom more than three-quarters were stationed in the Reserves, i.e. the small farming areas. The "average" Assistant Agricultural Officer was expected to cater to about 8,000 families. Under him the Agricultural Instructor was supposed to look after an average of about 1,200 farmers and the lowest-ranking Assistant Agricultural Instructor after some 500 farmers.

Animal health, husbandry and breeding is the responsibility of the veterinarians[1]. There are usually one Veterinary Officer (V.O.) and/or one Livestock Officer (L.O.) and one or two Assistant Veterinary Officers in a Veterinary District Office, and they are supported by several Veterinary Assisants and Veterinary Scouts.

The agricultural and veterinary field staff work in co-operation with other government departments and with the Local District Council, formerly called African District Council (A.D.C.), which is run by elected African members. The main source of finance for local schools, hospitals, water supplies etc. are cesses on the marketed agricultural produce. Originally it was thought essential that the African District Councils should finance a major part of the agricultural services, since agriculture is the main source of their revenue. With increasing agricultural

[1] The fact that animal husbandry was placed under the veterinarians led to a number of problems and frictions. See: Royal Commission Report p. 372.

activity, a system was introduced whereby a proportion of the basic agricultural services was paid for in the first place by the African District Councils and later on reimbursed by the Government. However, this system proved to be unsatisfactory in many respects. Since 1957 the staff and agricultural services have been paid directly by the Central Government. Local Councils continue to finance dipping schemes, Farmers' Training Centres, seed and stock farms and water supplies and to be responsible for minor irrigation schemes. However, the Central Government advances loans to Local Councils which are used for these services and which have to be repaid out of the Councils' revenues.

The various Government departments as well as the Local Councils send their representatives to a District Team, headed by the District Commissioner, which has (1) to establish local priorities, (2) to coordinate the work of different institutions and agencies, (3) to ensure the continuity of work, (4) to organise finance, (5) to present the plans to the Provincial Officers for approval and (6) finally to present them to the people of the District.

The funds for schemes outside the routine work of the local staff were allocated by the African Land Development Board (ALDEV) — the name changed frequently — whose members were nominated by different ministries. The Provincial Agricultural Committee, consisting of representatives of different ministries and nominated members, was responsible for (1) initiating plans; (2) preparing estimates of schemes; (3) submitting them to ALDEV; (4) carrying them out; (5) proposing appropriate by-laws to be passed by the African District Councils, and (6) ensuring consultation with African District Councils. The District Teams had to advise and assist the Provincial Committee. ALDEV ceased to exist in 1963 as a result of the new constitution which removed the distinction between the Non-Scheduled and Scheduled areas. The Board's loan-making functions have been transferred to the new Agricultural Finance Corporation and its power relating to grants to the Central Agricultural Board. The responsibility for carrying out such schemes lies with the Ministry of Agriculture, and the District Commissioner is responsible for the issuing and recovering of loans.

c) Methods of Approach

The methods used by the field staff since the mid-fifties deserve special attention. Extension work, as currently applied in the U.S.A. or in Western Europe, played but a minor role. There was decreasing direct pressure, as exercised during the campaign for soil conservation work before and during the Emergency; instead, the field staff worked within a framework of *general incentives* and *selective pressures*, whereby all possible "power-keys" of the administration were used in order to promote the introduction of improvements and innovations. In detail we can distinguish between the approach to (1) the farmer, (2) the location and its land tenure, (3) the activity, (4) the farming unit, and (5) aid measures.

α) Approach to the Farmer

As far as extension work — in the western sense of the word — was concerned, efforts were concentrated on promising locations and interested individuals with farms somewhat above the average. They were approached on visits, on Farmers' Days and at local shows. They were asked to group themselves into "Better Farmers' Clubs" and to attend Farmers' Training Centres [1]. The selective approach to the progressive individuals was supplemented by the general enforcement of rules and by-laws passed by the Local Councils or the Central Government. These rules and by-laws were to ensure that the soil was conserved and that good husbandry of cash crops was exercised. Those who violated the by-laws were reprimanded and, in a few cases, prosecuted and fined.

β) Approach to the Location and its Land Tenure

The British agricultural administration, after a decade of frustration, came to the conclusion that land tenure reform is a prerequisite to any sustained smallholder development. Wherever possible, either with pressure — as in Central Province during and after the Emergency — or with friendly persuasion — as among the Kisii and several Kalenjin tribes, where enclosures have been popular for several decades already — land consolidation, registration or enclosures were pushed ahead, followed by an intensification of the production and marketing development services. This really meant a concentration of efforts on (1) receptive tribes, (2) densely populated districts and (3) areas well suited for major cash crops.

The main advantages of consolidation and registration in Kenya are [2]: (1) security of investment in land, (2) protection against costly and time-consuming litigation, (3) access to credit against land mortgage, (4) the possibility to control tick-borne diseases on individual holdings, making it possible to keep exotic cattle, (5) a reduction of distances between house and fields, making it possible to apply manure with wheelbarrows. The main disadvantages are: (1) the appearance of a landless class [3], (2) loss of communal grazing for landless people, (3) the creation of a rigid structure in land tenure and (4) the substantial costs in terms of money, personnel and administrative change. It is noteworthy that land enclosures and consolidation have become rather popular. In some locations, prior to the arrival of the land survey team, the farmers have privately consolidated and demarcated their holdings. A council of elders works with the survey team and has to identify the already existing boundaries.

Within the process of agricultural development in Kenya a number of distinct stages in landownership can be recognized. Initially there is the traditional law

---

1 A Farmers' Training Centre is an agricultural school to cater for adult students, male or female, on a short series of courses on crop and animal husbandry. It operates under the direct supervision of the Agricultural Department and is financed by the A.D.C. A major function is the "in-service" training of the lower field staff. Most important, however, are short courses for farmers who attend with their local Agricultural Assistants. In 1963, Kenya was provided with 24 F.T.Cs. having a total of 1,267 beds. The work of the F.T.Cs. is generally considered to be very successful. Several of the F.T.Cs. have been built by aid from the USA.

2 For a detailed discussion of the land tenure reform in Kenya see bibliography.

3 Land reform did not make people landles who used to have rights on the land. Landless people existed before land reform, however, it brought them more clearly into the open.

and custom with individual rights to the land cultivated by one family, and communal rights on all grazings including the fallow. With increasing land shortage, people tend to individualize the grazing rights on and around their fields. The next step is the enclosure, i. e. the holding is surrounded by a fence or hedge. Finally the holding is registered — if necessary consolidated — and becomes a privately owned piece of land which can be sold or used as security for a loan.

γ) Approach to Activities

There was and still is no compulsion to grow certain crops in Kenya, a procedure one might expect in a country where the administration is the active force behind agricultural development. However, those who wanted to or could be persuaded to grow coffee, tea and pyrethrum, or to keep grade cattle, had to follow by-laws about husbandry standards laid down by the Agricultural District Council. Usually a whole bundle of

Table 4. *The economics of smallholder production of coffee in Nyeri district. Averages derived from the surveys made in 1962 and 1963—64*

Item	Age of Crop (years) 1	2	3	4	5
Hours worked per acre	1552	627	1494	1729	1669
Hired labour as *per cent* of total	73	57	67	48	62
Total labour costs in *Shillings* per acre[1]	294	137	277	424	407
Other expenses in *Shillings* per acre[2]	391	17	84	64	49
Total Cost *per acre*[3]	685	154	361	488	456
Cherry Sales, *shillings per acre*[4]	—	—	512	2251	4297
Balance, *shillings per acre*	—685	—154	151	1763	3841
Accounting profit, *shillings per acre*[5]	—	—	428	2187	4248

[1] Labour costs if all work was done by hired labour paid at the current rate.
[2] Seedlings or stumps, spray material, manure, mulch and fertiliser.
[3] Without land rent, interest, amortization of investment in trees etc.
[4] At 45 cts/lb.
[5] Sales minus other expenses. The accounting profit indicates the labour income created by the activity.

Source: Farm Econmic Unit: Some Economic Aspects of Agricultural Development in Nyeri District, 1962. Nakuru 1964. — Ditto: Economic Investigations of Farming in Nyeri District, 1963—64. Preliminary Results. Nyeri 1964, Mimeo.

complementary innovations was propagated as a unit. Thus cash crop planting was, as a matter of principle, put together with row-planting, terracing, mulching, fertilizing, etc. The lower field staff only to a limited degree had to deal with the difficult task of extension: more important was their supervision of those who grew cash crops. A case in point was the introduction of coffee and grade cattle into peasant farming [1].

Africans in Kenya — in particular in Central Province — had been under the influence of the exchange economy for a considerable time. There was an acute land shortage, and for some years there had been increasing pressure for permission to grow coffee and to earn cash. It was therefore no problem whatsoever to make people grow coffee. It was, however, a great problem to introduce coffee production in such a way that good husbandry was exercised and a high quality of produce obtained. Smallholder production of plantation crops or milk usually shows serious shortcomings: poor yields, poor quality, poor processing and poor prices. Kenya's Department of Agriculture endeavoured to avoid these pitfalls by trying to develop the industry step by step and to remain in full control of all plantings and all grade animals.

With *coffee* the procedure adopted was as follows:

— Research was undertaken as regards the appropriate locations and varieties. This was fairly simple because one could draw on the experience of neighbouring European Coffee estates.
— Nurseries were established, financed by the A.D.C.s and run by the field staff.
— Planting was supervised.
— Pulping factories initially run by the Department were established.
— The nurseries and factories were handed over to coffee growers' co-operatives which were supervised by the Cooperative Officer.

In order to obtain a licence the grower had to satisfy certain demands of the Department of Agriculture.

— The Agricultural Instructor would site the cash crop area on the holding and peg out the contour holes, the spacing, and the number of trees.
— The grower must maintain a sufficient acreage of mulching grass. The agricultural Instructor would indicate where the mulching grass should be planted.
— The farmer was allowed to grow only a certain variety of coffee. He had to buy the planting material from official nurseries. In this way they got priority allocations of the best varieties available.
— No couch grass was allowed on the coffee plot.
— The grower had to follow instructions as to soil conservation, planting, pruning, processing, and disposal. Most coffee plots had to be terraced.
— The grower must join the local coffee growers' cooperative society, the staff of which organised processing and sales.

In the beginning, most of the supervision had to be done by the agricultural field staff. In the course of time, however, more and more reponsibility was handed over to the cooperatives. It became one of their duties to see that their members followed the by-laws laid down by the A.D.C. Growers who did not abide by the regulations were not allowed to bring their cherries to the pulping fac-

---

[1] For tea se chapter II. The procedure of introducing pyrethrum in smallholdings in Tanzania, which followed the procedure developed in Kenya, has been described in some detail by the author in: Agricultural Development in Tanganyika. Berlin-Heidelberg-New York, 1964, p. 70.

tory for processing. Officers and instructors of the Department of Agriculture who used to be the supervisors changed in time to the role of technical advisers, providing farm plans and organising the distribution and control of small loans.

The drive for supervised coffee production proved to be remarkably successful. Thanks to comparatively good husbandry, high returns are obtained per acre and per hour of work. The Farm Economic Survey Unit has estimated the accounting profit per hour worked in 5-year-old coffee as 2/54 shs, while wages in the former Reserves are around 2/— shs for an eight hour day (see Table 4). The coffee acreage rose rapidly from almost nil in 1950 to 95,000 acres (official figure) in 1963 and most of it is showing a standard of husbandry which is closer to that of the plantation industry in Kenya than to that of smallholder coffee elsewhere in the world. According to an unconfirmed estimate in 1965 the actual acreage amounted to 121,500 acres, compared with 80,500 acres of plantation coffee. Almost 250,000 Africans were registered as members of coffee growers' co-operatives.

With *milk* as a cash enterprise, produced by grade cattle, the procedure was not as straightforward as with coffee or pyrethrum. It is rather interesting to reconstruct how trial and error led to today's successful procedure.

In the early fifties, farmers were urged to plant grass in order to reduce overgrazing and erosion damage. But they were rather slow in responding. The Agricultural Officer, Nyeri, reported: "Too often does one come across a patch of tall grass, and upon enquiring why it is not being used receives the answer that this is the Agricultural Department grass and is for viewing Officers only". At that time it was the Government's policy to encourage the keeping of native Zebu cattle and improved Zebu breeds in African areas and to select high milk-producing blood lines out of the indigenous stock. The Veterinarians considered the replacement of disease-resistant Zebu with high yielding but disease-prone Guernseys, Jerseys, Ayrshires and Friesians as too risky and complicated for African smallholders. Sahiwal bulls and artificial insemination schemes with Sahiwal semen were introduced in several high altitude areas. Dipping and spraying schemes were initiated. However, the smallholders could not imagine that any kind of Zebu cow was good for milk production and worth the trouble and costs of dipping or spraying. The break with traditional animal husbandry was not evident enough. Farmers started to buy exotic cattle on their own, rather ignoring or even in outright opposition to the Veterinary Officers.

Finally the Veterinary Department adopted a different policy and actively supported the introduction of exotic cattle in smallholdings. However, the introduction of grade cows into certain Reserves was made at a moderate pace in order to retain control of disease, to prevent unsuitable people from wasting valuable animals and because of cash shortage for loans and supervising personnel. Before anybody — in the relevant areas — was allowed to bring grade cattle onto the land, he had to obtain a "No Objection" certificate, ensuring that the animal was registered. The requirements for the certificate were that the farm had sufficient grass, that — in most cases — Rhodes Grass leys were established and that a suitable perimeter fence was available. For those who bought grade cattle on a government loan, conditions were even more stringent:

— the holding had to be boundary-fenced and paddocked,
— the owner to live on the holding,
— there had to be sufficient grazing or fodder grass,
— and inoculation and artificial insemination crush must be available.

In several locations artificial insemination was introduced. A Veterinary Scout, who had to inseminate the cows, was established in almost every village of the relevant area.

The Co-operative Officer, as is the usual procedure when introducing a new enterprise, tried to organise co-operative responsibility in the collection and marketing of the milk.

Table 5. *The economics of smallholder milk production with grade cattle in Nyeri district: 1963—64* [1]

Number of cows per farm	3
Production per cow per year	455 gallons
Per cent of production sold	62%
Per cent of production consumed in the household	24%
Per cent of production fed to calves	14%
Average price per gallon sold	2/08 shillings
Value of total milk production	950/—
Value of milk sold	581/—
Total expenses per cow per year excluding labour	59/—
Of which concentrate account for	22/—
minerals	6/—
fodder	10/—
Artificial insemination, drugs etc	21/—
Accounting profit per cow	891/—
Number of hours per cow per year	360
thereof feeding	187
milking	87
selling	24
treatment	25
fencing	4
washing utensils	22
other items	11
Accounting profit per hour worked	2/47 shillings
Proportion of work done by hired labour	45%

[1] Othaya, Mathira and North Tetu Locations only.

Source: Farm Economic Survey Unit: Economic Investigations of Farming in Nyeri District: 1963—1964, Preliminary Results. Nakuru-Nyeri 1965, Mimeo.

The movement for grade cattle on smallholdings followed that for coffee with a time-lag of some years (it rapidly gained momentum in the early sixties). The demand for milk was stimulated by the income from coffee; profits out of coffee farming were available for re-investment; this, as well as loans from aid giving agencies and from the Government, led to a remarkable build up of grade stock in African hands. The resettlement areas were stocked exclusively with grade cattle. In mid-1965 approx. 120,000 head of grade cattle (unconfirmed estimate) were owned by African farmers, of which roughly 50 per cent were kept in the former Reserves (according to unconfirmed information 40 per cent of the marketed milk in mid-1965 came from African farms). Three distinct phases in the development of milk production have thus become discernible:

1. Originally there were only poorly fed, low yielding Zebu cows in African areas.

2. In an intermediate phase more and more of the better farmers kept better fed and higher yielding grade and mixed grade cows. The disease risk is rather high in this intermediate phase, because Zebu cattle, neither sprayed nor dipped, graze in the immediate neighbourhood of grade cows.

3. In a final phase — reached in resettlement areas and aspired to in some very advanced districts in Central Province — all stock in the district is of grade or mixed grade blood and regularly sprayed or dipped. The native Zebu cattle, which used to graze on the roadside and were kept by traditionally-minded farmers and the landless, are transferred into grazing schemes elsewhere in accordance with the by-laws of the Local District Councils.

Table 5 provides some information about the profitability of dairy cattle in a smallholder setting. According to the estimates made in Nyeri District, where milk prices are high because of purchasing power created by coffee, the accounting profit per hour worked amounts to 2/47 shs, which is an amazingly high figure compared with any alternative use of labour except for coffee.

δ) Approach to the Farm

The Department of Agriculture considered that land tenure reform and supervised cash activities did no suffice to develop a sound system of farming. It therefore developed a follow-up process, which had three stages:

— A minimum standard layout was recommended for all farms, which involved (1) a basic soil conservation layout and (2) the siting of buildings so that they would not obstruct subsequent development.

— A simple farm layout was the next step, but only for those who applied for it. It involved
  — the siting of a homestead and vegetable garden area;
  — the siting and pegging out of an area for permanent crops such as coffee or tea;
  — the division of the cultivable parts of the farm into a number of fields which might be used for arable crops, or grass, or semi-permanent crops such as pyrethrum and pineapples.
  — the siting of fences, farm buildings and any necessary additional soil conservation measures.

— *The farm plan* was intended to make the fullest possible use of the holding by seeking to delineate the following areas:
  — Homestead, farm buildings and garden.
  — The permanent cash-crop area, which would already have been demarcated in the stages before, but might require some additional development or revision.
  — A number of unit-size fields, as nearly equal in size as possible, which would be rotated with crops and good-quality grass for the maintenance of dairy cattle.
  — Where necessary, trees sited on odd convenient corners, or in lines along hedgerows, to provide fuel and timber.

In the early sixties the field staff devoted much of its time to *simple farm layouts* and to *farm plans*. Those who were successfully growing cash crops were urged to participate and to invest part of their additional income from coffee into the realisation of the layout or plan. Pressure and incentives were used to obtain cooperation. In the Central Province, where compulsory villagisation was introduced as a security measure during the Emergency, only those who agreed to have their farm planned and to build permanent houses and who were willing to pay the nominal fee required for the plan were allowed to re-establish their homes on their holdings. Up to March 1963, 35,000 layouts had been made covering 394,000 acres and 5,234 plans covering 88,000 acres. These figures only refer to layouts and plans actually made by the Department of Agriculture staff. A far greater number of farmers have arrived at similar farming systems without having had any formal assistance. On the other hand, a greater number of the layouts and plans which were prepared have fallen into disuse.

ε) Approach to Means of Production

The Agriculture Department worked on the principle that no "gifts" should be made. Experience showed that the free provision of material and services was frequently misunderstood, leading to a waste of money and personnel and to a stifling of the initiative for self-help. Farmers had not only to pay for coffee plants, tea stumps and pyrethrum splits, but also for the licence to grow these crops. Dipping and artificial insemination had to be paid for; and a fee for any services in connection with a farm layout or plan. The revenue from these fees was nominal in relation to the actual costs, but the principle was considered essential.

The provision of loans is the necessary complement of a "no-gift" policy. The development of a newly consolidated holding and the introduction of cash crops brought a great demand for additional finance. The realisation of the farm plan of a five-acre holding might require investment in planting material, fencing, water supply, cattle sheds, grade cows, etc., amounting to between £ 50 and £ 100. In the beginning, some loans were granted by the A.D.C. and re-imbursed through ALDEV. Another source of credit has been that provided by the co-operative societies for their members. One of the principal features of the Swynnerton Plan was the provision of more credit to smallholders. American Aid (I.C.A.) established a revolving fund. In the early sixties, the International Bank for Reconstruction and Development and the Federal Republic of Germany made loan money available. Where land consolidation has been achieved, and/or where Africans can pledge a reliable income from other sources, some loans are given by private banks. Most of the money has to be repaid within five years at an interest rate of 5.5—6.5%.

Loans to smallholders proved to be a powerful instrument in the hands of the field staff. The loan was given for specific production development purposes and only at the recommendation of the Agricultural Officer, who, of course, used his position of power to further proper husbandry and endorsed the applications of those who followed his advice. The basic security for a short-term or medium-term loan was seen in the proper selection of applicants, careful planning, and the close supervision of the farmer's progress. This, of course, took much time. Nevertheless, in spite of the care in selection, a number of court proceedings had to be instituted for the recovery of outstanding loan repayments, and this took up a considerable proportion of the working hours of well paid government servants. However, much of the loan money was not recovered, partly because the British Administration, shortly before independence, shied away from unpopular measures.

2. Difficulties, Benefits and Costs

a) Difficulties

There is no doubt that the appeal of the late fifties to earn cash on consolidated holdings was much more effective and popular than the pressure for better soil conservation and better subsistence as had been exercised in the early fifties. Nevertheless, a great number of difficulties were encountered which were not so much of an "objective" or "technical" nature, but had their roots in the political, social and cultural setting of the "subjects" concerned, i.e. the smallholders and their European Government Officers. Large scale farming had provided most of the technical answers. The agricultural Department knew, more or less, what could profitably be grown on different soils and under different climatic conditions. Most of the difficulties were encountered when trying to get people to do what was obviously advantageous, already exercised by the better African farmers in the district and certainly known to most smallholders.

α) Socio-Political Handicaps

A major handicap was the lack of identity between government personnel and the people. Too many farmers for too long a time considered everything the European Agricultural or Veterinary Officer proposed as bad for no other reason than pure animosity towards colonial rule. Serveral politicians found it effective to exploit these sentiments and to encourage noncooperation. The agricultural services usually had to concentrate on the few interested farmers. In view of the great number of small holdings it was simply impossible to contact the mistrusting mass of subsistence farmers who, although knowing the advantages of some obviously profitable innovations, farmed according to the traditional pattern because the tribe was expecting conformity from its members. The spreading of innovations and improvements was certainly slower than need have been, because of racial tensions. Kenya under the British had to forego the push to extension which can be achieved if Agricultural Officers and political parties work together.

Another important feature was, and still is, the striking difference between the tribes. The Kikuyu, Wameru and Waembu not only showed remarkable activity and tenacity in fighting the British during the Mau Mau, but they also became one of the most active farming communities in Africa, south of the Sahara. It is interesting to note that another Bantu tribe, the Kisii of Nyanza Province, although not having had the pressure towards social change which the Emergency brought to the Kikuyu, is also progressing rapidly. The reaction of the Kalenjin tribes, the Elgeyo or Nandi for instance, to arable farming in enclosed holdings was also encouraging. Progress on the coast was slow, although the agricultural potential seems to be significant. The coast had the highest allocation of staff and funds after the Swynnerton plan of any province in the country, but the production response has been negligible. There are few places in East Africa where the British agricultural development efforts started as early, were pursued so consistently, and yielded such disappointing results as has been the case of the Waluo in Central Nyanza. The Swynnerton plan did not fare any better than pressure and persuasion had done in the mid-thirties or group farming in the early fifties. There was a fierce demand for loans in the late fifties, but the issuing of further loans had to be stopped in 1963, because repayments were 95 per cent in arrears compared with a national figure of 20 per cent. One of the main stumbling blocks of Luo agriculture has been the traditional system of land tenure, whereby land is communally owned and individually held. Efforts to introduce and enforce land consolidation in 1956 and 1957 were considered by the Waluo to be "colonial" and "a punitive measure". Luo attitude changed somewhat in 1964 and 1965. Political independence brought some degree of activity to Central Nyanza.

The lack of identity between the Government and the people together with political expediency, particularly after the Lancaster House Conference in 1961, handicapped the use of administrative authority. The Government, where it met reluctance, opposition and lethargy, refrained from pushing matters. The British administration in its final years shied away from unpopular measures, even where they were urgently needed, in order to overcome certain institutional obstacles such as traditional land tenure. However, some additional personnel and funds had to be made available in all districts, although not all of them met with the hoped for production response. The economic results of smallholder development consequently suffered.

The concentration of staff on the interested farmers with acreages somewhat above the average had certain advantages. However, the large number of small landholders remained almost completely untouched by extension efforts. Even in the more developed districts not more than around 10 per cent of the total farmers have really developed their holdings. The bulk of the land is still devoted to poorly husbanded subsistence crops. Another area which had to be neglected — in order to use the staff as economically as possible — was the production of food crops which remained at a low level of efficiency. A serious handicap to extension work was, and still is, the fact that the extension officers have to approach the man, while much or even most of the work is done by women who — according to custom — have to grow most of the food used in the household, who have the established right to organize their own affairs as they please, and, as they are largely illiterate, stick to traditional patterns of cultivation which in turn leads to a large acreage for subsistence while little land is left for the men to grow cash crops on [1].

β) Shortcomings in Husbandry

Another set of problems of no less importance than the socio-political one was posed by the shortcomings in husbandry. Experience shows that a significant proportion of the smallholders are money-conscious. Apparently, there is always a certain number of farmers who are willing to adopt profitable innovations. The introduction of proper husbandry, however, which is the key to sustained economic success, proved to be a much more difficult proposition. New crops are usually better husbanded than traditional ones because the farmer attaches great hopes to the economic return of the innovation. But attention usually slips after an initial push. There seems to be a discrepancy between the type and intensity of farming on the one hand, and the farming mentality on the other. The African cultural heritage does not include careful farming methods for intensive land use as is the case in many parts of Asia. The farming mentality of most of the African smallholders is still akin to the experience of generations who practised shifting cultivation, who learnt that the harvest depends on the acreage and the fertility of the land, who left the task of re-establishing soil fertility to the fallow, who were not "caretakers" but "users" of the land, who took the yield per acre and per cow as something given by God, and to whom the idea of consistently pushing up yields by careful and ever-improving farming practices was a completely strange notion. It is remarkable, considering the cultural heritage, that husbandry in Kenya has reached the present level. Nevertheless, progress has undoubtedly suffered because there has always been the tendency among smallholders to plant too much and to weed too late instead of cultivating a small piece of land properly; to start a new enterprise without doing it properly; and not to look after every item of the farm and the household regularly, carefully and efficiently. Again and again smallholder run into heavy losses because they lack the proper unterstanding and the drive to carry out adequately, over some length of time, what they are trying to do. These farmers put themselves into a vulnerable position when incurring consider-

[1] The Agricultural Department tried female instructors, female A.A.O.s, and tried training women at Farmers' Training Centres without getting very far.

able debts, for instance in stocking their land with grade cows before they are fully competent to keep them in milk and in good health. Few other agricultural activities are as demanding, in respect with husbandry, as dairy cattle in East Africa. Smallholders have done better than expected in controlling tick-borne disease. But much more economic weight must be attached to certain aspects of animal husbandry as for instance calf-rearing or the calving interval, which according to our estimates is about 18 months on the average. Farmers with poor husbandry constitute a danger to the agricultural development efforts in that failure at the intermediate stage may mean disillusion and reversion to mere subsistence farming [1].

In order to avoid these pitfalls of smallholder development, the Agricultural and Veterinary Departments tried to remain in close control of almost everything, every planting of coffee or tea, every grade cow, better farmer and his loan. Up until the Lancaster House Conference outright pressure was occasionally used on the negligent ones, and successfully so. Some farmers even had their coffee uprooted because it was not properly husbanded. Nevertheless, it will be a long time before husbandry in Kenya reaches the standard of Japan, Taiwan or Egypt. It is not so much the lack of understanding or knowledge — the information required can, as a rule, be obtained on the spot, because there is an Agricultural Assistant or Officer within easy reach of almost everybody — but the lack of implementing this knowledge properly was, and still is, one of the major bottlenecks.

γ) Inefficient Labour Use

Kenya is an underdeveloped country with extensive rural underemployment. Nevertheless, during the drive for cash enterprise after the mid-fifties, it appeared that labour is one of the limiting factors in production development. This is particularly evident in the case of late planting and weeding, with the corresponding loss in yields. When questioned, smallholders almost invariably reply that they have too much to do in the specific season.

Somewhat larger farmers in the former Reserve, who could intensify their land use by employing others, argue in unison that the labour costs are too high in relation to profit expectations and risks, although they have access to much cheaper labour than anybody else.

In a certain number of cases the low labour productivity per "man equivalent" can be explained by few working hours, i.e. the preference of leisure to additional income. However, many if not most of the farmers of Kenya are already beyond such an attitude. It can certainly be assumed that the actual working hours on Kenya smallholdings are higher than elsewhere in East Africa's peasant agriculture. Considering the vast acreage cultivated with the hoe and considering that it takes 200 to 400 hours to

---

1 See: MacArthur's report on Nandi farmers, 1964.

cultivate, plant, weed and harvest one acre with the hoe as the only tool, it can normally not be said that the phyiscal effort is insufficient. The explanation probably lies with the low labour efficiency during the working hours. Farmers and their helpers are obviously working unsystematically. Observing them, one gains the impression that most of the work could be achieved with less physical effort in a fraction of the time simply by using proper tools and an appropriate working rhythm. But repeated attempts to introduce better tools have almost always failed. Those who employ paid labour are usually not accustomed to organise the working day and to supervise their men. The extravagances in the use of labour by almost everybody who can afford to pay somebody for doing the hard chores on the holding came up as an important handicap in smallholder development. The experience of the British indicates, furthermore, that better tools and efficient labour use are much more difficult to introduce than new crops, better seeds or mineral fertilizer.

δ) Handicaps to Farm Planning

The British Agricultural Officer attached great importance to the preparation of farm layouts and farm plans and their application by the farmer. However, shortcomings in husbandry and labour use and in particular the presence of too many family relations on the holdings proved to be a major obstacle. Rather frequently the holding could not be considered a farming unit, because it had to be subdivided among different members of the larger family units, i.e. the wives, brothers and other relations, everybody attaching priority to subsistence needs. Experience showed that the full returns of smallholder development can only be attained if there is a constant drain of "surplus" family members to employment outside the holding, as well as better tools and higher efficiency in the use of labour.

ε) Handicaps of a Colonial Administration

Finally, attention has to be drawn to some difficulties which are unavoidable as long as smallholder development has to be carried out by expatriate officers. They have certainly been efficient administrators and technicians, and in Kenya, more so than anywhere else in East Africa. However, expatriate officers went on leave every $2^1/_2$ years. They were frequently transferred to other locations. In Nyeri and Central Nyanza, for instance, the average length of time spent by an Agricultural Officer in the district was not more than $1^1/_2$ years. There has been little continuity in supervising subordinate personnel. Most of the Agricultural Officers tried their best to identify themselves with African demands. However, they came from a different background and were living in a different world. The identity between the administrator or extension officer on the one hand and the peasant on the other, which is so important for the introduction of innovations or the execution of unpopular measures and which could be taken for granted in Europe, the USA or Japan, did not exist in colonial Kenya. Most of the officers worked with a remarkable degree of devotion in order to promote the welfare of the African peasant. But not all of them understood the different social and cultural position of those they wanted to help. For a very long time they underestimated the smallholders' capacity for commercial production.

b) Benefits and Public Costs. An Analysis in Three Districts

The agricultural administration and extension services, the aims, institutions, methods, and difficulties of which have been described briefly in the foregoing pages, brought important benefits to the rural population and the general economy, but involved substantial public expenditure. There seems to be a consensus of opinion that this work has been highly successful, but it is extremely difficult to substantiate this statement by quantitative evidence, i.e. to compare the benefits with the costs which had to be carried by the Local Councils, the Treasury and — mainly — British aid. Fairly reliable estimates about the sales of major cash crops are available. The public costs of smallholder development, however, are not readily available. The breakdown of the expenses of the Ministry of Agriculture does not provide sufficiently detailed information as to what the money was spent on and what was done by the personnel employed. It is obviously impossible to associate a certain percentage of the increase in marketed production to certain facets of the smallholder development policy, i. e. to land consolidation, extension, small loans etc. Until 1964 land consolidation, for instance, was handled at the provincial and district levels, and administrative officers worked part-time on consolidation along with other duties. The costs of all these services have to be considered "joint inputs".

In order to obtain an approximate idea about the relation between benefits and public costs of smallholder development, the available data in three districts, populated by three different tribes, have been collected and interpreted:

(1) the Kikuyu in Nyeri District and the Aberdares, (2) the Elgeyo and Marakwet, populating a district named after them, which stretches over the Western Rift Valley Escarpment, and (3) the Waluo of Central Nyanza, living close to Lake Victoria. It can be assumed that these districts are representative of the more important types of farming and the different farming mentalities prevalent in the high and medium potential areas of the former Reserves (see: Appendix IV).

a) Benefits

A general feature of the benefits which accrue from agricultural development services in Africa is their wide variation between different locations and tribes. Conditions in Nyeri, Elgyo-Marakwet and Central

Nyanza are a case in point. The three districts are endowed with varying conditions as regards soil, climate and markets — excellent (Nyeri), good (Elgeyo-Marakwet), and fair (Central Nyanza). The production responses, however, differ far more than is warranted by the objective factors alone. Marketed agricultural production 1952—1964 in Nyeri rose at a rate of 13 per cent per annum and 1954—1964 in Elgeyo-Marakwet at a rate of 16 per cent, while there was stagnation in Central Nyanza.

Table 6. *The value of marketed agricultural produce in three districts in £'000*

Year	Nyeri [2]	Elgeyo-Marakwet [3]	Central Nyanza [4]
1952	237	.	.
1953	141	.	.
1954	190	.	216
1955	155	.	206
1956	209	.	272
1957	203		194
1958	205	107	205
1959	241	102	189
1960	270	129	279
1961	324	140	245
1962	362	176	173
1963	454	275	187
1964 [5]	633	230	412

[1] Marketed production is defined as that which is sold outside the district. Since there is a cess on the produce, only part of the actual marketed production is reported.

[2] Without cattle, hides and skins.

[3] Without hides and skins.

[4] Without cattle, hides, skins, sisal, and bananas. Part of the maize production of Central Nyanza is illegally exported to Uganda, where higher prices are paid. These exports are not recorded. African areas only.

[5] Without settlement areas, added to the districts.

Source: Annual Reports of the Agricultural and Veterinary District Offices.

The figures of table 6 have, of course, to be interpreted with caution. The information about sales of maize, beans, vegetables, and bananas is hardly trustworthy. Since there is a local cess on the marketed produce, only part of the latter is reported. Some of Nyanza's maize is illegally exported to Uganda, where it fetches a better price. The major cash activities, however, (coffee, tea, pyrethrum, cotton, rice, milk, and wool) which make up the larger part of the marketed produce, are reported properly.

Doubtful as some aspects of the available figures might be, they are evidence enough to conclude that up to 1963 Kenya's economy hardly benefited from the efforts made in Central Nyanza. The only result was that it allowed the area to support a larger population at subsistence level, something which might have been achieved with hardly any government effort at all. There were some indications in 1964 and in 1965 that the farming mentality of the Waluo is changing and that the investments made during several decades of agricultural administration and extension may pay off in future years. However, during the period covered by the available information, the progress in farming was certainly very small. On the other hand, the benefits of the development in Nyeri and Elgeyo-Marakwet are manifold and substantial:

— There is evidence that the expenditure for services led to *improved subsistence* and nutrition and reduced the costs for famine relief. Judging by the export of hides and skins, meat consumption more than doubled in Elgeyo-Marakwet within 8 years. In coffee-growing areas, a high and sustained local demand for milk, at prices up to 3 shillings per gallon, is noticeable. A substantial proportion of the income from cash crops is invested in commercial dairy production. According to the estimates of the Farm Economic Survey Unit, approximately 20 per cent of the milk produced is consumed in the household.
— The *value of marketed production,* within the period given in Table 6, trebled in Nyeri and doubled in Elgeyo-Marakwet. As a result, substantial benefits accrued to the rural population.
— The *increase in family cash income* amounts to almost 100 per cent of the increase in marketed production .
— An *increasing proportion of the marketed produce is taken up by exportable products.* On the other hand, the use of imported means of production is far lower than in the large farm economy. Both these factors contribute to an improvement in the national balance of payments situation.
— In due course, there has developed an increasing *demand for non-agricultural products* and services. More consumer goods (textiles, bicycles, transistor radios) and more means of production are bought. According to the Agricultural District Book, the 1962/63 purchases of fertilizer and insectizides in Nyeri amounted to £ 19,000 compared with £ 5,000 in 1959/60 ... For both districts substantial sales of fencing materials, dairy equipment and hand tools are reported.
— There is also a *higher demand for those agricultural products* which are produced elsewhere in the district or country: sugar, tobacco, cotton, meat, milk etc. This in turn gives more scope for cash activities in areas not suited to exportable products. A case in point is the trebling of cattle sales in Central Province since the early fifties.
— The increase in marketed produce together with the increase in demand for goods produced elsewhere leads to *substantial revenues.* In the first place, there are the cesses of the Local District Councils. In 1961 these amounted to £ 15,600 in Nyeri and £ 2,100 in Elgeyo-Marakwet. Secondly, there are those revenues derived from import levies on such goods as transistor radios and sewing machines, which go directly to the Treasury. Thirdly, there are the revenues received from the increasing processing and transportation activities in these areas.
— Furthermore, there has been *substantial capital formation.* In Nyeri, for instance, 13,500 acres of coffee and 1,760 acres of tea were established by 1964. The costs of establishing one acre of peasant coffee amout — according to the estimates of the Farm Economic Survey Unit — to 1,500 shillings and that

of peasant tea to 2,500 shillings. These small plantations therefore represent a total capital investment of about £ 1.2 million. Were these investments to be valued at the cost usually incurred by the estates (i. e. including labour charges), it would lead to a figure several times higher. Capital formation through the replacement of Zebu cattle with grade cattle amounts to at least £ 100,000 in Nyeri and £ 50,000 in Elgeyo-Marakwet. In addition there has been widespread capital formation in farm buildings, terracing, fencing, cattle dips, water supplies etc. The greater part of the investment made in recent years must have come from private sources, i.e. the income of rural households.

— Another important aspect is the *improved employment situation.* It is a remarkable feature of agricultural development in East African smallholder communities that a substantial proportion of the work connected with cash activities is performed by paid labour. The small agricultural entrepreneur who is the backbone of progress in both districts apparently prefers to employ casual labour instead of fully exploiting the work capacity of himself and his family. Furthermore, a significant number of farms — and rather frequently the better ones — seem to be owned and run by absentee landowners, working in towns as teachers, government officials etc. who employ labour to do the chores on the holding. Estimates made by MACARTHUR of Nyeri indicate that approximately half of the work in coffee and tea is done by hired labour. In 1963 there were 13,500 acres of coffee in Nyeri, with, according to MACARTHUR, a labour requirement of 1,500—1,900 hours each. If these estimates are correct, total paid employment in coffee alone would amount to 5,000—6,000 full-time labourers [1].

— The situation as regards *soil conservation improved significantly.* The change in land tenure (consolidation and enclosures) and the growing interest in cash activities made more and more farmers "land conscious". After the completion of land consolidation, farmers have shown a willingness to bench and terrace their land. Land tenure reform und supervision of good husbandry of cash crops did more to prevent soil erosion than the pressure for terracing in the early fifties.

— Finally, there has arisen a *greater awareness* of the income potential *of cash crop farming.* Though subsistence still appears to be the main concern of the majority, more and more farmers with some business sense are making real efforts to face up to the demands of a changing society. There has been a remarkable build-up of indigenous commercial and managerial experiences, particularly in Nyeri where the number of cooperatives increased from 8 in 1952 to 40 in 1960.

— Some of the better farmers in Nyeri and in Elgeyo-Marakwet have become settlers or bought large farms in the former White Highlands. As a result of cash crop farming on smallholdings, they have gained at least some experience

[1] There were 6,250 more families than land units in Nyeri in 1962. This figure corresponds rather closely with the estimate above. According to BARBER the employment effect of smallholder development in Nyeri District has not been as high as originally expected. It is certainly true that so far only a fraction of the land is as intensively used as is economically advisable, i. e. the Nyeri District still has a great potential to absorb additional man-power productively. Although with the present information, it is not possible to estimate accurately the employmen effect of smallholder development, evidence in Nyeri seems to indicate that the effect on unemployed and underemployed rural labour is higher with inputs in smallholder development than with any alternative use of public funds.

as agricultural entrepreneurs. Thus we can conclude that there has been a substantial formation of *"human capital"* in the form of knowledge, experience and more progressive attitudes towards economic activities.

β) Public Costs

The total costs for agricultural administration and extension can be conveniently divided into two sections:

— the overheads made up by the Ministry in Nairobi, the Provincial Office, research etc. and
— those occuring at the district level.

While there is no logical method of distributing part of the overhead costs to a specific segment of the agricultural economy, we can, however, reconstruct the costs at the district level.

The main items of agricultural expenditure in the above-mentioned districts were the following:

— Salaries for Agricultural Officers (4—8 officers in Nyeri, 2—4 in Elgeyo-Marakwet, 7—11 in Central Nyanza).
— Personal emoluments for Technical Assistants (there being 5—10 Assistants to each Officer). Up to 1957 these emoluments had to be paid by the African District Council, which was reimbursed for part of the outlay through the Central Government. Thereafter the payments were carried by the Central Government directly.
— The District Office's expenditure for vehicles, travelling, minor employees, and casual labour.
— Funds for specific schemes such as soil conservation, land consolidation, tea development, dippig schemes, loans, etc. net of fees and repayments.
— Costs of the Veterinary Office, including salaries for 2—4 Officers and several Technical Assistants, expenditure for transport, medical supplies, etc.

Table No. 7 gives an estimate of these costs in the three districts. The data have to be used with caution, particularly those concerning the early fifties. The information available was not complete and missing links had to be supplied on the basis of the preceding and following years. We can, however, assume that the facts are reliable enough to inform us about the magnitude involved. As can be seen from Table 7, the public costs for agricultural administration, extension, land consolidation, and minor schemes in the progressive District of Nyeri, with a rural population of 256,000, hovered around £ 50,000 per annum. In the less densely populated Elgeyo-Marakwet District with its 161,000 people, costs started at £ 12,000 in 1952, reaching double and nearly three times that amount between 1955 and 1961 when loans of more than £ 40,000 were distributed. Costs in Central Nyanza, with a rural population of 639,000 but with an agricultural situation singularly lacking in progress, were somewhat lower than in Nyeri.

A significant feature of this kind of development expenditure lies in the fact that approximately 80 per cent of the total costs were for personal emoluments, either in the form of staff salaries or wages for minor employees and casual labour. The remaining 20 per cent also has a high wage content, since most of it is spent on construction projects, repair of vehicles, or minor irrigation works.

Table 7. *An estimate of public costs for agricultural development in three districts* [1]

Year	Nyeri [2] £'000	Elgeyo-Marakwet [3] £'000	Central Nyanza [4] £'000
1952	40	12	27
1953	41	14	28
1954	77	18	30
1955	52	18	40
1956	55	20	42
1957	54	23	45
1958	46	23	60
1959	48	28	50
1960	46	30	49
1961	50	26	55
1962	49	17	54
1963	50	18	47
Population [5] in 1962 in '000	256	161	639

[1] Local expenditure, according to the Vote Book of the Agricultural and Veterinary District Office, plus agricultural expenditure of the African District Councils. The estimate includes Dara, Aldev and Swynnerton-Plan funds for general betterment, land consolidation, soil conservation, minor irrigation schemes, loans and tea-development (without factory). It excludes investments in water supplies. Loan repayments and fees are deducted. Where information was incomplete, estimates have been made. The costs of staff have been estimated as follows: Number of AO's, AAO's and Technical Assistants multiplied by a medium salary for the relevant year, plus 30 per cent for fringe benefits (leave, pension, housing).

[2] The estimates for Nyeri include expenditure for famine relief labour, soil conservation work during the Emergency, and local expenditure for land consolidation.

[3] The estimates for Elgeyo-Marakwet include general betterment funds amounting to £ 22,000, loans amounting to £ 40,000, and minor irrigation work amounting to £ 10,000.

[4] The estimates for Central Nyanza include expenditure for minor irrigation, particularly in 1958.

[5] Without township population.

The activities of the Agricultural and Veterinary Departments were consequently greatly influenced by higher wages and salaries. The £ 50,000 spent in 1964 represent far less activity than the same amount spent in 1954, when per capita emoluments were lower. Land consolidation in Nyeri, which was done in the midfifties with comparatively cheap labour and personnel, would have been far more expensive under the conditions of 1964. Therefore, the figures of Table 7, although being fairly constant over a time, represent different levels of activity, with a definite peak between 1955 and 1961 when most of the Swynnerton-Plan funds were spent.

3. Evaluation

The range of benefits which cash activities have brought to those districts where people cooperated is impressive. There is no doubt that the objectives — as stated before — have been attained to a remarkable degree. The success of Nyeri and Elgeyo-Marakwet is by no means exceptional in Kenya. Many other districts show similar rates of progress, and the performance in other districts around Lake Victoria is significantly better than that in Central Nyanza.

The social and political consequences of smallholder progress are obvious in the Kenya scene. Taking what seems to be practical out of socialism and capitalism, the Kenya African National Union (KANU) is accepting smallholder farming as appropriate to the present stage of development, and is actively promoting enclosures and consolidation leading to "absolute ownership" of land in the hands of smallholders. KANU is thus taking a rather different attitude towards smallholder farming from that of TANU in Tanzania, which aims at large scale communal or cooperative farms. Here no enclosures or consolidation of smallholdings take place, although such a move would certainly be welcomed by the people of some areas.

The achievement of the technical objectives has also been outstanding. It has been shown that the problem of rural Kenya does not lie in overcrowding but in poor land use; that small farms properly managed are producing better incomes than would any other occupation for an unqualified man. Cash farming, as done under the supervision of the Agricultural Department, has definitely reduced erosion. Kenya is one of the few places in the tropics where plantation crops and exotic stock have been introduced into the indigenous farming system with a level of husbandry which — although being lower than that in the large farm economy — is far higher than the usual standard. Up to now only a small fraction, perhaps 5 per cent, of the high potential land in the Reserve is farmed up to its economic potential; but significant nuclei of modern farming have been established.

However, impressive as the achievements are, they had to be met by heavy expenditure. Although expenditure has been rather low per farm, it has been high per 100 acres and in relation to the marketed produce. Local costs in our three case-studies — Nyeri, Elgeyo-Marakwet and Central Nyanza — amount to 15—18 per cent of the value of the marketed produce. If we relate the funds spent for agricultural development to the consequent additional marketed produce, we get a relationship of 1 : 2 in Nyeri, 1 : 3 in Elgeyo-Marakwet, and 1 : 0 in Central Nyanza. That is to say, each pound spent on agricultural administration and services realised £ 2, 3 or 0 in additional marketed produce. The picture is more favourable if we take the more recent years. Relating the public costs of 1958—1961 to the marketed production of 1960—1964 — assuming a time lag of three years — we find that each pound spent resulted in a £ 4 increase in marketed produce over the period 1958—1960 in Nyeri, a £ 3.5 increase in Elgeyo-Marakwet and £ 0.5 increase in Central Nyanza.

Table 8. *An economic appraisal of agricultural administration and extension in three districts*

	Nyeri [1]	Elgeyo-Marakwet [2]	Central Nyanza [1]
1. Marketed produce sold in the district, £'000 per annum	306	166	236
2. Total local expenditure for agr. and vet. services, £'000 per annum	51	24	43
3. Expenditure in per cent of marketed produce, 1955—64	17 per cent	15 per cent	18 per cent
4. Expenditure for agr. and vet. services per holding and per annum in £	1.1	2.4	0.4
5. Expenditure for agr. and vet. services per 100 acre farmland and per annum in £	25	..	5
6. Additional £ marketed produce for each £ local expenditure over a 10 year period [3]	2	3	nil
7. Additional £ marketed produce for each £ local expenditure over the last 4 years [4]	4	3.5	0.5
8. Cesses paid on the marketed [5] produce, £'000 per annum	10.6	1.8	11.1
9. Revenue to the Treasury £'000 per annum [6]	36.7	19.9	28.3
10. Total (8 plus 9)	47.3	21.7	39.4
11. Balance (10 minus 2)	—3.7	—2.3	—3.6

[1] Average of marketed produce 1955—1964 and of local expenditure for agricultural and veterinary services 1952—1961 with a time-lag of three years.

[2] Average of marketed produce 1958—1964 and of local expenditure for agricultural and veterinary services 1955—1961.

[3] In Nyeri and Central Nyanza: marketed produce 1955—1964 surpassing that of 1952—1954, divided by expenditure of 1952—1960. In Elgeyo-Marakwet: marketed produce 1960—1964 surpassing that of 1958—1959, divided by the expenditure of 1955—1961.

[4] Marketed produce in 1960—1964 surpassing that of 1958—1960, divided by expenditure 1958—1961.

[5] Market cess going to the District Council.

[6] Based on the arbitrary assumption that 12 per cent of the value of the marketed produce is taken by taxes on goods bought by the farmers, i. e. can be considered as revenue coming from primary income.

Thus, the figures of Table 8 indicate that, up to now, smallholder development has been of doubtful profitability to the Treasury itself. If we add the cesses paid to local councils to the revenues which the Treasury is likely to get out of the increase in primary income and deduct from this the local public costs, we find that all three districts show a slight negative balance. In addition the overheads for agricultural administration, research and education have to be considered.

The calculations in Table 8 have to be supplemented by numerous considerations. We have to take into account that:

— the Treasury not only receives revenues out of the primary income created. The revenues out of the secondary income might be even more important;
— the estimates do not include the increase in subsistence production and in produce which is marketed locally. The Treasury is likely to reap some benefits from the reduced need for famine relief and the local trade.
— the situation will improve considerably when more of the coffee and tea planted and the animals bought come into full production;
— the repayments of loans in the years to come, particularly in Elgeyo-Marakwet, will make the picture more promising.

Yet it must be asked whether it is justifiable to regard the total increase in the marketed production as the consequence solely of public inputs. There are a number of cases in African farming where smallholders communities commercialised rapidly with only a few and cheap services provided: cotton in Sukumaland, Robusta coffee in Uganda, cashew nuts on the East African coast, groundnuts in West Africa, etc. Some production increase was bound to take place in Kenya's Reserves, even if only a fraction of the actual services had been provided:

— Let us first look at fertile Nyeri. The compulsory return of all Kikuyu to Central Province during the Emergency, the resulting increase in land pressure, the "fierce desire" — according to Humphrey in 1944 — of the Kikuyu to make money in order to buy land, the rapid social change brought about by the Emergency and many other factors prevalent in the Kikuyu scene would, in any case, have led to economic repercussions. Coffee prices were high and people wanted to grow coffee. If the administration had simply relaxed the rules controlling the coffee industry in the Reserves and had told the peasantry to go ahead, plantings of Arabica coffee would perhaps have been even more extensive than today.
— The indigenous drive in Elgeyo-Marakwet would probably have been weaker than in Nyeri. But some development at any rate would have taken place there, too. The increase in marketed production of European potatoes and the establishment of mechanised wheat production are likely to have occurred with little help of the Agricultural Department. There are other cases in East Africa where Africans established medium-sized wheat farms with a minimum of public support, for instance in the Arusha-Mbulu districts of Tanzania.

"What would have happened if the smallholders had had to rely solely on their own initiative?" Although this question might be justified, we can nevertheless say that important and far-reaching benefits are definitely attributable to the efforts of the Agricultural and Veterinary Departments. For example:

— Land tenure reform which is proving to be the very basis of sustained smallholder development, particularly in the sphere of animal husbandry, goes to the credit of the Government's efforts.
— Without the field staff's persistent persuasion and close supervision there would be little smallholder tea growing of economic relevance, and probably little smallholder pyrethrum [1]. The quality of husbandry in coffee production, yields per acre, and income per grower would probably be much lower.
— The fine hedging, terracing, and grass strips planted on the contour, which are to be seen on a number of farms, must go to the credit of the Agricultural Department. Widespread models for the expansion path of the farming systems have been established.
— With fewer public services, the introduction of exotic stock, which might be considered to be the most important move for future farming in Kenya, would have been much slower and might have broken down because of disease or disorganised marketing.

The above list of some of the important factors which have to be considered when judging the economics of agricultural services in the Reserves poses for us the difficult problem of how to make a balanced assessment of the role of these services. In attempting to do this, we are inclined to give decisive credit to public efforts. Although we hesitate to attribute all the increase in marketed produce to public funds, we do not hesitate to say that the investments in smallholder farming which have been made, if properly sustained — and this can be done fairly cheaply — will yield substantial returns for many years to come.

A number of models and principles for successful smallholder promotion have been developed in Kenya which are likely to stimulate economic progress elsewhere in Africa. It is true that the expenditure has been much higher than could possibly have been financed out of the revenues of an underdeveloped economy. Smallholder development in Kenya was largely paid for by aid provided by Great Britain. However, there were some factors operating in the Kenya scene which kept costs high and returns low and which are not necessarily part of the British approach.

The following points should be considered in regard to the high costs:

— A rather costly period of trial and error had to be passed before it was realized that land tenure reform and promotion of cash activities are the "spearheads" of rural progress, bringing other improvements in their wake. Too much personnel and too many funds were used for too long a time for objectives which found little response among the peasantry: soil conservation work, better subsistence, improvement of Zebu cattle.

---

[1] Large farms have been growing pyrethrum for a long time without neighbouring smallholders being induced to take up production, as was the case with coffee in many places. Historical evidence indicates that the field staff had to provide the first incentive and the initial organisation for marketing.

— With agricultural services it is necessary to sustain efforts continually, provided that the people continue to respond. This was difficult to realize. Again and again the Treasury had to cut down the allocations without warning and the Agricultural Department was not allowed, for obvious political reasons, to cut the allocations for non-responding areas in favour of those where people did respond. A pro rata cut was generally enforced, which had most discouraging effects.
— The Agricultural Officers — in the period we cover — were mainly expatriates who had to be provided with lengthy home leaves, free passages, cheap housing and pension rights. Rural extension services, with salaries higher than those that are paid in industrialized societies, can hardly be economic in smallholder communities of underdeveloped societies. An earlier but gradual Africanisation would have made the services probably somewhat less efficient but much cheaper and thus — perhaps — more economic.

But for these factors, which kept costs high, and the aforementioned difficulties keeping returns lower than they need have been, the relationship between public costs and the increase in marketed production would probably have been much better. An efficiently run agricultural administration and extension service, working with the full support of the political party and backed by sufficient Government authority, using the British strategy of smallholder development —

— land tenure reform,
— concentration on the most rewarding cash activities, locations and tribes,
— organisation of supplementary services (nurseries, markets cooperatives),
— close supervision of key activities in their initial stages (plantation crops, grade cattle, use of loans, cooperatives),
— no general compulsion, but the judicious use of a system of pressure and incentives,
— follow-up of progress in cash activities by farm layouts —

is likely to succeed in the future Kenya as well as in other African countries, provided — and this provision is of key importance — that the services are efficiently organised.

II. Smallholder Tea Development (Field Sector Only)

Within the general efforts to introduce plantation crops into peasant agriculture, special programmes for tea, pineapples etc. came into existence. Most interesting and promising has been the approach to smallholder tea, the economic results of which — because of its long term nature — are not yet available. Up to now it can be assessed only on its initial results.

The first tea in Kenya was planted in the 1920's. Subsequently an important tea industry, mainly owned by large commercial companies, developed around Kericho, Nairobi, and on the Nandi Hills. It had generally been held that tea growing on a smallholder basis was uneconomical because of the difficulties in leaf transport and processing. Attitudes changed in the 1950's. Experimental planting was started by the Department of Agriculture and showed promising results. By 1960 a total of 1,500 acres had been planted and properly husbanded. In June 1964 the estimated acreage of 10,658 acres was divided among 19,775 growers, giving an average tea plot of .54 acres per grower [1]. Smallholder tea development has, since

Table 9. *Capitalisation and economic return of smallholder tea development (Planning estimates for a 25,000 acres programme including existing plantings and plantings up to 1968/69)*

A. Investments	£
(i) Field Sector[1]:	
International Development Association	1,000,000
Commonwealth Development Corporation	900,000
Kreditanstalt für Wiederaufbau	212,000
Kenya Government	20,000
Former Marketing Board Loan[2]	27,000
KTDA Revenues[3]	1,941,000
Sub-Total	4,100,000
(ii) Factories (Provisional)	
Commonwealth Development Corporation	1,230,000
Commercial Companies (direct and through KTDA)	740,000
Government (including about £ 100,000 Ragati)[4]	350,000
Sub-Total	2,320,000
(iii) Government Sector	
Housing and Offices (including about £ 30,000 spent)	90,000
Roads (including about £ 100,000 spent)	1,550,000
Sub-Total	1,640,000
TOTAL	8,060,000
B. Costs per acre	
Field Sector, approx.	164
Total	320
C. Value of tea to be produced in 1975/76[5]	5,100,000

[1] Public costs only. 20—40 p. c. of the actual costs of tea establishment on the fields are financed by the growers.
[2] Used to establish tea prior to 1961, when the First Plan started.
[3] Revenue from plantings already in production.
[4] Investment prior to 1961 in the Ragati factory.
[5] Assuming an average yield of 1,000 lbs. made tea per acre at an average price of Sh. 4/— per lb.

Source: The Operations and Development Plans of the Kenya Tea Development Authority, Nairobi, 1964.

[1] The actual acreage per grower is probably somewhat higher and the number of growers smaller than is given above, because growers tend to increase their acreage by having shadow partners.

its inception, proved to be so popular and successful that it has been able to attract large sums of international finance for future expansion.

The First Plan for the planting of 11,000 acres was drawn up in 1961 and the necessary finance was obtained from the Commonwealth Develop-

"Smallholders Tea Growing Areas in Kenya"

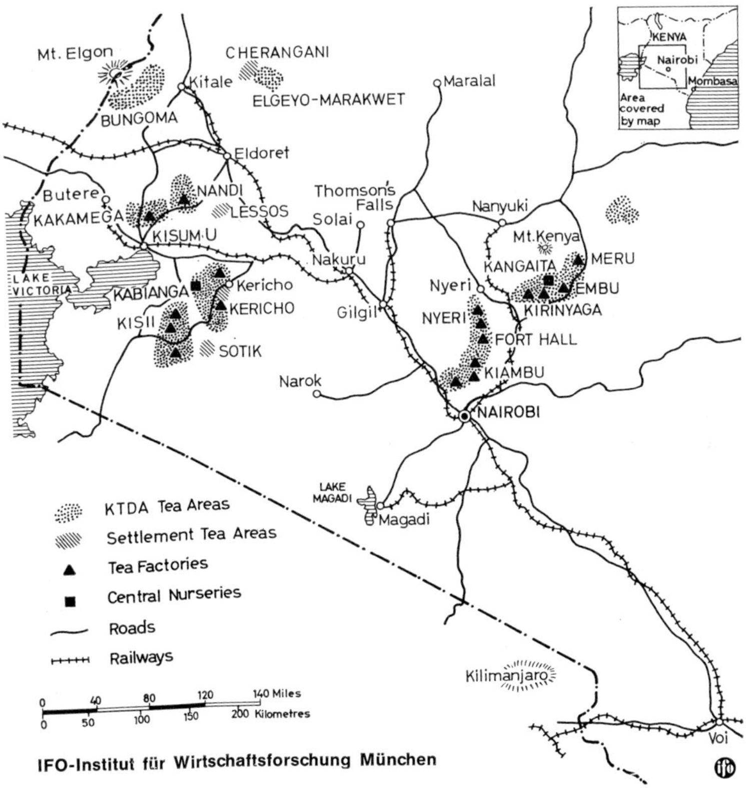

ment Corporation and the Federal Republic of Germany. During 1964 a further loan was obtained from the International Development Association (an affiliate of the IBRD) which will enable a Second Plan involving 14,400 acres to be put into effect by 1970.

1. Aims, Institutions and Methods

a) Aims

The aims of smallholder tea development have largely been the same as those of cash crop promotion in the Reserves in general. Of special relevance to tea are the following:

— A high yielding cash crop is needed in the densely populated high altitude areas which are unsuitable for coffee.
— Tea will bring diversification of cash activities into districts which depend almost entirely on coffee.
— Tea is a reasonably reliable crop bringing a fairly high element of stability into the economy.

b) Institutions

The early experimental developments in Nyeri and Kericho Districts were administered by officers of the Department of Agriculture and financed by the Kenya Government through two statuary boards, the Central Province African Grown Tea Marketing Board and the Nyanza and Rift Valley Province Tea Marketing Board. Government loans were made to these Boards to finance expenditure and to finance the first factory at Ragati, Nyeri District. This was opened in 1957. In 1960 the Special Crop Development Authority was established, which continued to be responsible for smallholder tea until succeeded in 1964 by the Kenya Tea Development Authority (KTDA), a statuary corporate body responsible to the Minister for Agriculture.

The whole complex of smallholder tea development under the Authority's auspices is designed to be a self-financing entity. The KTDA is financed partly by loans raised on commercial terms and partly by self-generating revenues — from cesses levied on growers. The cess income also provides revenues for the redemption of loans. Factories are separately financed units. It is argued that it would be much better if the factories could be run as cooperatives like the coffee pulping factories where growers could participate in the framing of the factory policy and control would be entirely African. But tea processing is on such a large scale and is relatively so complicated that at this stage the cooperative has proved impracticable.

The KTDA has its own staff for headquarters and leaf collection. The field supervision is carried out by officers and instructors employed and paid by the Department of Agriculture, who are on full-time secondment to the Authority. The entire cost of the field staff is reimbursed to the Government by the KTDA. The tea development in Settlement areas (eventually to be about 2,412 acres) is not financed by the KTDA but by the Settlement Fund Trustees (see V. 1 b). However, the KTDA ist acting as managing agent in the four settlement schemes where tea is being planted.

The growers sell their leaf to the Authority which makes corresponding payments to them less certain cesses. The Authority subsequently sells the leaf to various factories, the period of ownership by the KTDA being the few hours between the time of purchase at the buying centres and the time of sale at the factory door. The Authority can thus be considered the trading intermediary between the grower and the factory.

The growers elect their own tea committee at divisional, district and provincial levels. It is the task of these committees to channel the suggestions and grievances of the growers to the Authority and to popularise the Authority's policy among the growers.

c) Methods

Smallholder tea is grown in twelve administrative districts. The areas that are generally suitable for tea have been demarcated and no tea is allowed outside the so-called "tea line". Farmers within the "tea-line", having consolidated or enclosed holdings, may apply for a licence to plant tea and for a loan for stumps. The question as to whether the tea should be established in large blocks, each grower having a garden within the block, or in small patches on the individual's holding, was decided at the very beginning: the opposition of the growers to the block system was such that already existing blocks had to be uprooted. Hence it has been the policy of the Authority that small patches of tea are developed on individual holdings in order to obtain the full interest and cooperation of the growers. This pattern has led to the widely scattered nature of the Authority's operations.

It is the Authority's aim to achieve an average of one acre per holding, a size generally considered to be the maximum within the working capacity of a family unit which does not employ extra labour. Planning is based on 1,000 pounds of made tea, i. e. 4,500 pounds of green leaf per acre in the sixth year after planting, the conversion factor being 4,5 pounds of green leaf to 1 pound of made tea. The envisaged average scope of tea production is thus 1,000 pounds per holding per annum.

The Authority produces stumps in its own nurseries or, where necessary, by purchase from commercial enterprises. These stumps are sold to the growers at 30 cents each. For the first 3,000 stumps, sufficient for planting approximately one acre, a grower may receive a credit of 18 cents a stump. A deposit of at least 12 cents per stump has, therefore, to be paid. No credit is allowed for purchases after the first 3,000 stumps.

Field supervision is carried out by officers and instructors who are responsible for (1) nurseries, (2) distribution of planting material, (3) the organisation of a credit scheme for nitrogenous fertilizer, (4) supervision of the maintenance of standards of tea husbandry set out in the Tea Cultivation Order, (5) general tea extension, (6) the issuing of licences to growers, (7) the maintenance of records, and (8) the administration of leaf collection and inspection where no leaf officer has yet been appointed.

So far the KTDA has strictly adhered to the principle of "close supervision" of all plantings. Serveral times a year the farmer is visited by an Agricultural Instructor, who has to fill in a form reporting the

— number of new stumps planted,

— number of infills,
— number of stumps that have died.

Each time he gives points to the performance with regard to: — planting, — fertilizing, — shadow, — mulching, — refilling, — soil conservation, — weeding, — pruning, — plucking table, — quality in plucking, and health of plants.

After the visit, the completed form has to be signed by the farmer. The evaluation of each visit is registered. At the end of the year the KTDA is informed about the farmer's acreage and the standard of his husbandry.

The leaf collection operations are supervised by Leaf Officers, assisted by subordinate staff, who are directly employed by the Authority. The tea leaf is delivered by head porterage to buying centres, where, after inspection, it is purchased. It is the Authority's policy to accept only "two leaves and a bud" in order to ensure quality and thus high prices for the processed tea.

The buying centres are sited so that the grower will not have to carry the leaf more than 1,5 miles. After purchase, the leaf is conveyed to the factory for processing. Ideally each buying centre will be situated no more than 10 miles from one of the factories processing smallholder tea.

At present the growers receive 40 cents per pound of green leaf, in quarterly payments. In addition there is a bonus at the end of the year according to the prices received at the auctions for the made tea. Before payment the Authority automatically recovers two cesses which are imposed on the green leaf produced by the growers:

— the revenue cess of ten cents per pound of green leaf, which is calculated to be sufficient to meet the continuing costs of the Authority when all loans have been repaid;
— The capital cess of 7 cents per pound of green leaf which is calculated to meet the costs of the grower's loan for the purchase of the stumps and development expenditure (feeder roads in particular), incurred by the KTDA. Growers who took no credit on their stumps purchases are still liable to pay the capital cess of 7 cents. However, they are paying off their obligations in a shorter time than those who bought their stumps on credit.

A credit for nitrogenous fertilizer is also recovered from growers by deductions against the leaf payments.

2. Difficulties, Benefits and Costs

a) Difficulties

Smallholder tea development in Kenya has worked amazingly well, considering the necessity for a number of rather complicated arrangements. The scheme is popular. There are only a few specific problems related to the current programme:

— The supply of stumps has been insufficient. The Authority has therefore had to set up two large centrally operated nurseries.

- In some locations growers try to sun-dry their leaf and sell the produce on the black market, thus avoiding payments of cesses and loan instalments.
- The husbandry of smallholder tea has been remarkably good so far. There are, however, indications that with increasing acreages yields tend to fall off, due, no doubt, to mounting demands on the field staff. Particular damage is done by the lack of weeding and by the intercropping of immature stands with maize and other demanding crops.
- With the latter point in mind, there is some doubt as to whether the planned target of 1,000 pounds of made tea per acre of mature tea will be reached. Since both cesses depend on the amount of green leaf produced, it will perhaps take somewhat longer than planned to recuperate the investments made. Model calculations indicate that the contingency provisions of the tea scheme are sufficient to absorb a drop in the average yield to 950 lbs per acre of mature tea. A yield of 900 lbs per acre would require changes in the financial arrangements, i. e. extra capital or a reduced rate of loan repayments. If the yields fell to 800 lbs per acre of mature tea the scheme would not be viable at the present cess rate. These estimates show the paramount importance of good husbandry on the part of the growers and of close supervision of each grower by the KTDA officers.

b) Benefits

According to the estimates in Table 9, after 1976, when all plantings have matured, smallholder tea development is expected to yield an annual production of £ 5.5 million, valued at present prices. If the tea prices remain as they are and if the KTDA is allowed to continue to work as smoothly as it has done so far, then it is likely that the target will be reached and that Kenya will reap the political, social and economic benefits which should accrue from this. (With regard to the nature of the benefits, see: C. I. 2. a.)

In areas with mature tea, some benefits can be seen already. The grower is receiving 40 cents per pound. Thus far experience shows that in good locations a second payment of 10 cents per pound can be expected. This adds up 50 cents per pound for green leaf at the buying centre. From this the revenue cess of 10 cents and the capital cess of 7 cents are deducted, leaving 23 cents as a minimum price where no second payment is made and 33 cents as the higher net receipt to the producer. Assuming an average yield of 4,500 pounds per acre of green leaf, the income of the grower will amount to shs: 1,035 or shs: 1,485 respectively. After the repayment of the capital loan, which will take 15 years for a grower who has taken maximum credit facilities, the price received will be 7 cents higher, giving a net receipt of up to 1,800 shillings per acre.

Table 10 shows some tentative estimates on the economics of tea production from the standpoint of the grower. It takes him roughly 8,200 hours to establish one acre of 5 year old tea. Expenses are made for the purchase of the stumps and for hired labour. The yield of 3,600 pounds in the fifth year indicates that the target of 4,500 pounds in the sixth year is likely to be achieved. The sales figures inform us that the producer actually received 33 cents per pound. In the fifth year, after deducting the expenses for hired labour, the profit amounts to roughly shs: 800 per acre. The performance of the sixth year and the years thereafter is likely to be even better. A comparison with Table 4 shows that coffee production is four times as profitable as the production of tea, which is, therefore, in the eyes of the farmers, a much poorer choice. However, the alternative is open only to those limited areas suited as well for coffee as for tea.

Table 10. *The economics of smallholder tea production in Nyeri district, 1963—64, per acre*

	Year of Maturity 1	2	3	4	5
Hours worked	2655	918	665	2006	1944
Hired labour in per cent of total	70	47	31	13	59
Total labour costs in shillings[1]	735	205	164	471	376
Other expenses[2] in shillings	1092	.	.	.	.
Total costs[3] shillings	1817	205	164	471	376
Yield per acre, pounds of green leaf	—	—	—	777	3614
Leaf sales in shillings[4]	—	—	—	257	1193
Balance in shillings	—1817	—205	—164	—214	817

[1] Labour costs if all work were done by hired labour paid at the current rate. Family labour costs are calculated at —/25 cents per M/E hours worked.
[2] Tea stumps.
[3] Without land rent, interest and amortization.
[4] 33 cts/lb of green leaf.

Source: Farm Economic Survey Unit: Economic Investigation of Farming in Nyeri District, 1963—64. Preliminary Results. Nyeri-Nakuru 1965, Mimeo.

The high percentage of the work performed by hired labour indicates that the idea of a self-contained family unit, growing tea for cash and food crops for home consumption, is illusory. It is apparently the small entrepreneur who is accepting the opportunity offered by the Authority's tea programme.

c) Costs

The costs of the two current plans covering the planting of 25,500 acres are estimated at £ 8,060,000 including all overheads, factories, staff housing, and road development. Of this, almost £ 2 million is expected to be carried by the revenue gained from cesses on already established tea. The only item not included in the cost estimate is the overseas inducement paid by the British Government to expatriate officers working on the scheme.

The costs of the Field Sector are assumed to amount to £ 164 per acre, and those of the total to £ 320 per acre. These estimates indicate that it is probably cheaper for the Government of Kenya to establish smallholder tea than plantation tea, for recent experience in the tea industry in Kenya shows that the costs of a new plantation work out at approximately £ 400 per acre.

There are a few other relatively unimportant items which might be considered as costs or losses. Tea is taking away land from alternative uses. However, much if not most of the land planted with tea was previously only extensive grazing, low yielding wattle trees, or even bush. In the same way, labour is being drawn into tea. But alternative employment of the same scope in the relevant high altitude areas hardly exists, and the only point of some weight that can be made is that pyrethrum cultivation has suffered because it was at the price level in 1964 certainly not competitive once there was the chance to use the available labour on tea.

3. Evaluation

Smallholder tea development which proved to be a failure in India, Ceylon and Java has made a most promising start in Kenya. The problems of collecting, inspecting and transporting the tea seem to have been solved satisfactorily. The tea produced has been of very high quality, fetching excellent prices on the Nairobi and London markets. The scheme has been so popular that applications for licences exceed the acreage that can be handled. Owing to the Authority's monopoly in the buying of the tea leaf a promising approach for the loan repayments has been made. Only the future will show whether KTDA can continue to work as smoothly as it has done so far and whether the estimates as to yields, costs and loan repayments can indeed be reached. The present observer cannot but conclude that the chances are good.

III. Grazing Schemes [1]

The approach of the Agricultural Department has been largely limited to the high potential areas receiving at least 30 inches of rain annually. In the semi-arid areas — where pastoral tribes, constituting a population of one and a half million people and covering 166,000 square miles, possess one third of Kenya's stock resources — development efforts focussed on "grazing schemes". These might be described as the institutionalised and ratio-

[1] Of the area receiving less than 30″, only about 38,000 square miles are in the 20—30 inch range and may be described as potentially good ranching country. Of this, but 4,500 square miles are taken by European ranches. 73,000 square miles, exclusively in the hands of pastoral people, receive 10—20″ and may be described as poor or marginal ranching country. About 60,000 square miles receive less than 10″ and count as virtually useless semi-desert.

nalized use of communally held grazings and individually held stock. In 1960, the area covered by organised grazing schemes amounted to 9.1 million acres.

1. Aims, Institutions and Methods

The aims of the grazing schemes were the same as those of smallholder development elsewhere. However, there was a difference in emphasis:

- The semi-dry areas are notorious for bad land use leading to erosion. The situation in Baringo District, for instance, has been depicted by L. H. Brown as follows: "Baringo District has reached an 'overgrazing end point' where most of the grass and the topsoil has already gone over large stretches of country and the ground is blanketed with thornbush, largely useless to man and beast alike, which cannot be eradicated without the expenditure of large sums of money . . . Land in a 35—40 inch rainfall area, once capable perhaps of supporting a stock unit to 4—5 acres, is now scarcely capable of carrying a stock unit to 20 acres . . . In the drier part of the district with a 20—25 inch rainfall, land once capable of supporting a stock unit to 10 acres is now scarcely capable of supporting one to 30—40 . . . Baringo is a case where the human population, in an attempt to maintain enough stock for their needs, have already to a large extent destroyed their own habitat." (p.16) Grazing schemes were considered to be the proper way to prevent erosion, to heal the damage already done, and to increase the carrying capacity of the land.
- The contribution of the semi-dry areas to Kenya's marketed production has been negligible. £ 1.7 million of marketed produce come from the 4,500 square miles of European ranches, while the 166,000 square miles used by the pastoral tribes contributed only £ 1.5, mainly from the sale of skins and hides. It is estimated that the actual gross returns in semi-dry areas are only one third of what they could be. In areas used by herdsmen the value of gross production per acre varies between —/50 cts and 7/— shs. In commercial ranching areas it varies from 1/75 to 20/— shs with up to 40/— on the most favourable range.
- The semi-dry areas have been a liability to the country's economy. In almost every dry year famine relief has had to be provided. The whole cost of famine relief in 1961—62 amounted to over £ 5 million, of which £ 3.5 million came from the U.S.A.; and most of it was for the people living in these areas. Apart from keeping the people alive, the money was entirely wasted from the point of view of economic development. The grazing schemes were intended to secure a better distribution of income between wet and dry years.

The main fault in the old system was the combination of communal range with unrestricted individual ownership of cattle, because neither the community nor the individual concerned themselves over the effects of their actions on the land. Two basic steps needed to be taken:

- limiting the number of stock to one the area can carry;
- controlling the movement of stock within any given area through rotational grazing systems.

Without these two basic practices, no investment — whether additional water supplies, tsetse clearing, reseeding, bush control or veterinary measures — had proved to be of the slightest value. Too often bush had been cleared, dams put in and cattle dipped, only for the inhabitants to repudiate agreements made and for the areas concerned to be seriously damaged by increasing numbers of stock.

A really bad case of overgrazing and erosion damage usually touched off Government activity. However, before funds were spent it was made clear to the cattle owners that they had to submit to appropriate District Council by-laws or Government regulations. As far as possible the provision of water was used as a bait. The Range Officer approached the cattle owners, arguing, "You say you want water here and here. All right, the Goverment will try to provide water at these points, but on condition that you accept grazing management. If anybody does not accept, let him go elsewhere". This in due course led to the so called "squeeze". Once a grazing scheme was set up, resulting in the reduction of stock, some cattle had to move to adjoining areas where overgrazing and erosion increased. This in turn left hardly any other choice to neighbouring cattle holders, but to ask for an extension of the area covered by grazing schemes (HENNINGS, 1961).

Where persuasion did not work and where increasing erosion required immediate action, Government regulations were imposed. In almost all cases the schemes were under the control of a Government Officer who commanded grazing guards and, if necessary, asked for the support of the tribal police. It was realised, however, that it would be difficult, for any length of time, to control the grazing routine, to keep stock numbers within the prescribed limits, and to enforce the dipping of cattle by compulsion alone. The Range Officers, therefore, from the very beginning tried to get the support of a "grazing committee" representing local stock owners of repute. The ultimate aim was to transfer the responsibility of the whole scheme to the grazing committee. Very few schemes actually reached this stage. In some exceptionally favourable cases it had been tried to establish or foster "grazing associations", wherein shares were allocated to the members according to the number of stock held by any individual. The grazing associations were supposed to raise their own funds, to ask for loans and to employ a manager.

Thus the typical grazing scheme included the following features:

— The demarcation of a certain area as a grazing scheme.
— The adjustment of stock to the carrying capacity of the land within the given area.
— The organisation of a culling programme and stock disposal.
— The castration of scrub bulls.
— The introduction of rotational grazing — simple alternation or irregular rotation in most, and regular rotation in some of the better schemes.
— The construction of dips and the provision of veterinary supervision.
— Provision of water through the construction of dams and wells.
— The extension of grazing through bush clearing and tsetse eradication where necessary.
— The provision of loans for the investments mentioned above.

— The payment of a fee of between 6 and 16 shillings per head of cattle per annum. The fee was collected by the District Council which had to service the loans.
— The supervision of the management rules by grazing guards and a Government Officer.

2. Difficulties, Benefits and Costs

Most of the grazing schemes ran into so many difficulties that they had to be abandoned. The emphasis was on administrative action in order to prevent erosion. Priority was given to areas which had been seriously misused. Grazing schemes were established not so much where they would give the best and quickest return, but where they were most urgently required to remedy a desperate situation. Thus land rehabilitation measures took precedence over improved animal husbandry. The first step taken was usually a compulsory reduction of stock numbers in order to re-establish the grass cover. The reduction of stock numbers, in turn, met particularly fierce opposition for a number of reasons, among which the following seem to be of paramount importance:

— Many of the pastoral tribes were already living on a submarginal level. The reduction in stock numbers at first reduced the supply of milk and meat for the household. The grazing schemes were usually imposed on the most denuded locations, i. e. where a comparatively long time was required to reap the benefits of the culling programme. The herdsmen and their families apparently did not have the means to bridge the time between the reduction in stock and the higher return in milk and meat as a result of it.
— Pastoral societies are built on an intricate system of human bonds established by the lending, renting, exchanging, and presenting of cattle for different reasons in different situations, the bride price being only one of them. Culling programmes simply do not fit into the traditional social pattern.
— The limitations of stock numbers under the rules meant that each family herd had to be split up into two — some animals were in the quota on the scheme and the remainder had to find their grazing elsewhere. As a result, the herdsmen were separated from their families and very naturally the idea was unpopular.
— The reduction in stock numbers put the greatest relative strain on the man with a lot of cattle, who was frequently an open or „hidden“ leader of the clan or tribe. Therefore, grazing schemes were often opposed by those who had the say in their communities.

Another weakness of most of the grazing schemes was that it involved a large collection of heterogeneous people with conflicting economic interests. The schemes apparently were too large with the responsibility resting on the District Council to which the herdsmen felt little allegiance. The collection of fees per head of cattle for the provision of water and dips usually was a task which the District Council could not handle.

In the fifties, i.e. during the initial years of the grazing schemes, most difficulties were overcome by Government authority. Where the schemes existed over some time a considerable measure of success was achieved. Supervised grazing management obviously led to the regeneration of good grasses, usually in a shorter time than expected, to an increase in the carrying capacity of the land, to healthy cattle, and thus to higher production and income. However, discipline slipped after the Lancaster House Conference and was never re-established; and the disastrous drought and flood of 1960 to 1961 made it difficult to follow any rotational grazing system. The combined effect of these factors was to reduce the area under any sort of grazing control from 9,1 million acres in 1960 to about 800,000 acres in 1963. In consequence, the £ 812,000 invested and the work of the staff, whose salaries are not included in the above figure, have largely dissipated. The only recognisable benefits are the survival of the roughly 1 million acres under control and — perhaps even more important — a push towards the change of mind which, since 1964, is taking place among some pastoral tribes.

The establishment of a grazing scheme involved non-recurrent (survey, demarcation of boundaries, roads, water supplies, staff housing, cattle dips, transport, etc.) and recurrent (wages of grazing guards, operation of water supplies, general maintenance etc.) expenditure. Part of the finance for the non-recurrent expenditure was provided by the sale of stock in a culling programme. The rest was covered by Government loans given to the District Council. It was expected that the District Councils would collect sufficient fees to repay the loans and to carry the recurrent expenditure, except for the Government personnel employed in the scheme. According to L. H. Brown, the average costs — excluding staff salaries — amounted to £ 57 per square mile or 1/60 shillings per acre, that is £ 800 per 10,000 acre scheme.

3. Evaluation

There was, and still is, an urgent need for the introduction of sound systems of grassland-use in the semi-dry areas. Most of the grazing schemes, as organized by the British, were technically sound, economically promising, and pliable enough for any social adjustments which might have been necessary. They failed or could not be re-established after the drought and flood of 1960 and 1961 only to a minor degree, mainly because they were used as political pawns in the struggle for independence and because the Government did not wield enough authority after the Lancaster House Conference. The British were probably right in exerting a certain degree of pressure, but they did not choose the right time for doing so. It is, after all, the duty of every responsible administration to use authority where people are obviously destroying their habitats and thus the nation's natural resources and where friendly persuasion does not show any results. If the move for grazing schemes had started several decades earlier, when colonial authority was still unchallenged, we would probably have firmly established grazing institutions over most of the good ranging land in Kenya today.

IV. Settlement in New Areas

1. Aims, Institutions and Methods

a) Aims

In addition to the efforts to improve existing smallholdings and to reorganize the use of grassland, which absorbed most of the funds and personnel, a number of supplementary settlement schemes on rain-fed land were carried out. Most of them were initiated at a time when the Department of Agriculture was taken with the idea that the African lands were overcrowded and overstocked and that there could be no improvement in farming unless a large number of people and cattle were moved to new land. This led to the search for unused land of acceptable fertility and to the establishment of settlements on it. A number of other objectives were considered as well. J. D. MacArthur lists the following[1]:

— Control of tsetse invasion and its eradication from certain areas.
— Exploitation of undeveloped land.
— Political land exchange between tribal and racial groups.
— Population transfers for administrative convenience.
— Settlement on the land of landless, unemployed families to provide them with means of supporting themselves.

b) Institutions

Finance came from the African Land Development Board (Aldev). The Provincial Agricultural Committee, consisting of representatives of different ministries and nominated members, had to (1) initiate plans, (2) prepare estimates, (3) submit them to Aldev, (4) carry them out, (5) propose appropriate by-laws to be passed by the African District Council, and (6) ensure consultation with the African District Council. The settlement became a distinct land unit with one or more officers in charge who were usually responsible to the Provincial Agricultural Office and in some cases directly to Aldev.

c) Methods

As regards the methods of settlement, three basically different types can be distinguished.

α) Settlement in Expansion Areas

The least expensive kind of settlement was the occupation of land in expansion areas, carried out by the neighbouring peasantry. Unselected families moved into the unused areas, cleared the bush and started to cultivate according to traditional patterns. The Government gave support by building feeder-roads and providing water (wells, hafirs). An example of

---

[1] MacArthur, J. D.: Economic Aspects of Land Settlement in Kenya. (Outline of Work) Nakuru 1965, Mimeo.

this kind of settlement is the Chepalungu area, taken by the Kipsigis (see Table 11). Public inputs per capita were low. In the case of Chepalungu they amounted to £ 4 per family. This simple type of settlement which has been of great importance in Tanzania and responsible for much of the in-

Table 11. *The costs of settlement schemes in the former Reserve (Aldev schemes)*

Time[1]	Location	Tribe	Altitude '000 ft.	Rainfall inches	Costs[2] £'000	Acreage[3] '000 acres	Number of[4] families settled	Costs per family
Expansion areas								
1949—58	Chepa-lungu	Kipsigis	8—9	40—60	27	71	6,800	4
Planned unirrigated settlements								
1939—63	Gede	Giriama	sea level	40	65	26	700	93
1949—61	Makueni	Wakamba	3—4	20—30	335	50	2,187	151
1949—62	Olen-gurone	Wan-derobo	8	40—60	25	15	289	88
1949—63	Shimba Hills	...	1	53	181	42	824	220
1950—59	Lambwe	Waluo	4	50	72	79	533	135
1951—58	Kimulot	Kipsigis	7	70	20	7	258	76
1954—60	Kaimosi	Nandi	6	40—60	11	6	181	63
1954—60	Itembe	Kipsigis	6	50	23	10	312	75
1955—59	Sarora	Nandi	6	40—60	14	6	202	69
1957—61	Lembus	Tugen	6	60—70	13	20	346	37
Irrigation Settlements								
1954—64	Mwea[5]	Kikuyu	4	25	806	6	1,340	601
1954—64	Perkerra[5]	Tugen	.	.	435	1.5	349	125

[1] Period in which the costs occurred.
[2] Loans and grants. Puplic inputs minus settlement fees usually without salaries for settlement officers.
[3] Total area, only partly usable.
[4] Families present at the end of the time given.
[5] See Table 12.

Source: African Land Development in Kenya, 1946—62. MACARTHUR, J. D.: Several papers about Settlement Schemes, MS. Nakuru, 1964.

crease in smallholder production — particularly of cotton in the Lake Region — had but few possibilities in Kenya (Coast Province, South Nyanza, West Pokot etc.) Most smallholder areas are surrounded by dry steppe country, forest reserves or large farming areas.

β) Planned Settlement on Rain-fed Land

More important have been the planned settlements on rain-fed land[1]. They were established in different parts of Kenya and with different

[1] The methods have been rather similar to those used by the British agricultural administration in Tanganyika (see RUTHENBERG p. 78).

tribes. The largest project, Makueni, in Machakos District, is situated in a low rainfall area close to the Wakamba Mountains. The Gede and Shimba schemes are to be found in the Coast Province and the Lambwe Scheme close to Lake Victoria. The remaining schemes, mostly of smaller acreage, are situated in high rainfall — high altitude areas. The methods employed differed widely. In most cases detailed plans for land use had been worked out. The settlers were placed on distinct holdings and usually provided with free rations up to the first harvest. In the earlier schemes the settlers received little or no loans and no part of the development costs was retrieved from the settlers. In the more recent schemes more loans were provided and loan repayments, rent charges and fees were charged to the settler. A proper selection of settlers has usually not been possible, the only criterion being that they had to be from the local tribe. The agriculturists who were in charge of the scheme virtually had to accept any landless families sent to them by their administrative colleagues. Rules were usually imposed covering soil conservation, husbandry, farm development, stocking with cattle, and residency of the settler on the scheme. However, these rules could be enforced in exceptional cases only, because of lack of personnel, lack of "power keys", lack of settlers, and lack of authority.

γ) Irrigation Settlement

Finally, there has been irrigation settlement, usually involving heavy investments in water and land. Minor irrigation projects, being too small to carry the overheads of a proper scheme management, were taken over by the local peasantry with the District Council in charge of the financial responsibilities. Usually the standards of water use, land preparations and plant husbandry remained at a low level and the economic return of the investment made by Aldev was meagre. (See Aldev Reports.) In addition to several of these minor projects, three major irrigation settlements have been set up (i) Mwea-Tebere in Central Province, where rice is grown by the Kikuyu and Waembu, (ii) Perkerra in Rift Valley Province, where onions, maize and beans are grown by the Njemps and Tugen and (iii) Galole at the Tana River, where cotton is grown mainly by the Pokoma.

In all three cases almost from the very beginning the land was given to the settlers, and a landlord — tenant system was established, with the Ministry of Agriculture as the benevolent public landlord. The schemes are governed by the Trust Land (Irrigation Areas) Rules 1962. On receipt of his licence of occupancy the tenant signs an undertaking to observe the rules. The duration of tenure is one year, and the licence is automatically renewed, provided that the tenant honours his obligations. Only one of the three schemes, Mwea-Tebere, is fully operational. The Perkerra scheme is run on a "caretaker" basis because of the human problems involved with the settlement of pastoral people as irrigation farmers, and the Galole scheme has still to be considered as a pilot project.

The organisation of irrigation farming at Mwea-Tebere is depicted by E. G. GIGLIOLI as follows: "Tenants are allocated four one acre plots as a rice holding. Only one crop of transplanted rice is grown per year ... No rotation is practised ... The rules are very comprehensive ... (and) ... particularly detailed on matters relating to cropping practices, water control, absentee ownership, and disposal of the crop. Management has three disciplinary tools to deal with recalcitrants, these are, in order of increasing severity: written warnings, prosecution and termination of licence ... Selection (of tenants) has been limited to the disinherited, and the Clan Committee have a very human tendency to unload onto the Settlement the less desirable members of their community ... (Nevertheless) in the three years that the Irrigation Rules have been in force only 19 tenants have had their licence terminated (Mwea section only). The basic organisational unit ... is a block of 2,400 acres of irrigated rice, supporting 600 families ... The Assistant Agricultural Officer is in charge of all activities on the block ... The day to day supervision ... is in the hands of the Field Assitants. Each man is responsible for 150 tenants ... All water control is in the hands of the Settlement staff. A tenant is not allowed to interfere in any way with the water regime in his field ... the Head Water Guard and his Guards are responsible for execution ... The last link in the chain between the Manager and the tenants is constituted by the Head Cultivator ... a tenant selected for his farming and leadership qualities ... The introduction of mechanical cultivation has made possible a degree of planning and discipline and extension unthought of in the past. The orderly progression of cultural operations made possible by mechanisation of puddling has produced an atmosphere in which strict discipline can be enforced without opposition ... It is now possible to meet every group of tenants ... before the beginning of each stage of the cultivation and to indoctrinate them on the measures. Repetitive indoctrination made possible by mechanisation is showing its effectiveness by both the increase in yield and the even distribution of the increase over all the tenants on the Settlement."

2. Difficulties, Benefits and Costs

a) Settlement on Rain-fed Land

Most of the settlements on rain-fed land concerned land which was marginal or difficult as regards rainfall, climate or soil, but which, provided it was farmed properly, could produce a reasonable income. There is hardly any scheme without its master farmer, whose fields show the economic potential of the area. However, the areas concerned proved to be too difficult and the period of trial and error so costly that the economic returns to the public investment in settlement remained negligible: The major problems have been the following:

— A high priced cash crop which is essential to sustain high settlement expenditure was generally lacking.

— Much was spent on items like housing, feeder-roads, water supplies, without finding the expected economic response among the beneficiaries.

— Many of the settlers were uninterested, landless people without any particular farming qualifications. The returns per family were not high enough to attract worthwhile settlers or to induce the settlers to do the required chores. Squatters preferred to remain where they were established rather than to move to a settlement scheme where they were subject to rules, rent payments and exhortations to develop the land.

— The Settlement Agency usually aimed too high and tried to do too much too soon with insufficient personnel. Some efforts were made to introduce rotational land use, to bring cattle into areas infested by tsetse, to use ox-drawn equipment, without a closer examination of the question as to whether these innovations were economically and technically feasible, and with insufficient finance to give the settlers a sound start on a more intensive farming system than the traditional one.

— There was evidently a lack of energy to press the settlers. Rarely did they conform to laid-down rules of good husbandry. They seldom turned out for communal work and — in most cases — did not pay their fees. Where pressure was exerted, it lacked consistency. In several cases it proved to be impossible to enforce the settlement rules, although they had been signed and agreed to in advance by the applicants. In other cases the schemes were opened for settlement before the lease and development conditions had been fully worked out. The settlers were not conversant with the conditions or did not know them when the plots were allocated and consequently did not feel obliged to follow them.

— The Aldev Report of 1955 comments: „Money for nothing is seldom valued and several projects have been pauperized by too much charity. A scheme which is planned on commercial lines, with an annual loan charge as a kind of burden, will be run more economically than a scheme which relies on free grants".

— The number and quality of the staff were usually inadequate to ensure that rules of good husbandry were obeyed. The economic dilemma is obvious: either the schemes had enough staff and the costs of staff became out of proportion to the returns; or there were not enough staff to supervise the settlers properly and to ensure sound farming, which again led to a poor performance.

— There was a lack of continuity. Some of the schemes which now are hardly recognisable as such might have had some measure of success if the same Settlement Officers would have stayed on long enough and if a reasonable policy had been pursued consistently over many years.

In the schemes listed in Table 11 not more than 5,800 families were settled. The local costs were £ 760,000, that is to say £ 312 per family. However, these costs do not include:

— salaries — at least not in most cases — of the Settlement Officers, which are likely to have been the most costly item of all;

— overheads of the settlement agency, the Provincial Office etc;

— the heavy burden of interest charges since substantial sums were tied up for years, even decades, with hardly a return.

Systematic surveys about yields and returns of these settlements have not yet been made available. They are likely to vary widely according to the peculiar circumstances of each case. While the Shimba scheme is generally considered a success, the Makueni and Lambwe schemes are thought to have been failures up to now. It is generally assumed that land use, yields and returns of settlement farms are no better than those of farms elsewhere in the former Reserves where similar conditions prevail. The funds invested in the careful planning as well as the extension activity of the Settlement staff have had few noteworthy results.

Table 12. *Costs and returns of irrigation development, large schemes only, in £'000*

	Mwea-Tebere Irrigation Rec.	Capit.	Total	Rev.[2]	Bal.	Perkerra Irrigation Rec.	Capit.	Total	Rev.	Bal.	Tana River Irrigation Rec.	Capit.	Total.	Rev.	Bal.	Grand Total Exp.	Rev.	Bal.
1954—56	35	138	173			22	88	110	.	—110	8	32	40	.	— 40			
1956—57	79	112	191	107	—588	16	33	49	.	— 49	6	16	22	.	— 22	1284	107	—1177
1957—58	80	106	186			22	55	77	.	— 77	24	104	128	.	—128			
1958—59	80	65	145			16	33	49	.	— 49	73	37	110	.	—110			
1959—60	81	229	310	74	—236	31	10	41	.	— 41	83	92	175	.	—175	526	74	— 452
1960—61	85	31	116	78	— 38	16	3	19	2	— 17	43	5	48	7	— 41	183	87	— 96
1961—62	80	8	88	105	— 17	14	3	17	2	— 15	41	24	65	.	— 65	170	107	— 63
1962—63	77	20	97	114	+ 17	20	7	27	2	— 25	.	.	41	19	— 22	165	135	— 30
1963—64	.	.	104	126	+ 22	.	.	46	13	— 33	.	.	44	14	— 30	194	153	— 41
Total	.	.	1410	604	—806	.	.	435	19	—416	.	.	673	40	—633	2512	663	—1849

[1] Rec. = Recurrent expenditure. Capit. = Capital expenditure. Rev. = Revenue from water rate. Bal. = Total expenditure minus revenues. Exp. = Recurrent and capital expenditure of all three schemes.

[2] Mwea also contributes in the last years £ 12,000 to the County Councils as a cess on paddy and £ 25,000 per annum towards a rice Stabilization Fund. The latter in late 1965 amounted to some £ 125,000 and has never been used.

Source: Ministry of Agriculture, Nairobi.

If we assume that the average costs per family, excluding overheads and interest charges but including salaries for Officers, amounted to £ 150 per family, and if we further assume that each family produces additional products worth £ 25, which is likely to be an optimistic estimate, we obtain a relationship between settlement investment and additional annual marketed produce of only 6 : 1.

It should not be overlooked, however, that some of the settlement schemes — Makueni for instance — have served as a nucleus for the expansion of peasant cultivation in surrounding areas, thus contributing indirectly to the opening up of hitherto unused land.

b) Irrigation Settlement

There is certainly no irrigation scheme which does not run into a number of technical difficulties in the use of water, the cropping system, the use of difficult soils situated as islands in good irrigation land, the choice of proper equipment etc. Even the most fortunate and successful, the Mwea-Tebere scheme, had its share of them. However, the major problems of irrigation settlement seem to be of a managerial and socio-political nature. Mwea-Tebere shows the organisational pattern which was aspired to in Perkerra and Galole, too, but which could not materialize because of numerous difficulties:

— The schemes had to be settled on the spot although it would have been advisable to run them for some length of time as a large farm in order to develop the most profitable farming system.

— In addition, political influence was exerted in the choice of settlers. They had to come from the local tribe, although it is obvious that settlers of some of the tribes with long experience in commercialized arable farming would make better irrigation tenants than people with a pastoral background or from backward tribes.

— Furthermore, it proved to be impossible to pick out the more promising applicants from the local population. The irrigation settlements more or less had to take those who were sent to them by the relevant local committees.

— The basic difference in the performance of irrigation settlements, provided land, water and crops are adequate, is due to the continuity and quality of management. The performance of management again not depends only on the personality of the manager, but also on the atmosphere he is working in and in particular the political backing he is receiving from the relevant tribal and government authorities. Owing to the premature settlement and the lack of proper selection of tenants, it became virtually impossible to enforce proper husbandry, which, of course, led to poor yields per acre. A significant proportion of the irrigated fields at Perkerra and Galole are full of weeds.

— Finally, attention needs to be drawn to the fact that tenants ask for irrigated land in order to grow their subsistence food, maize and beans in particular. This again leads to the absurd situation that some of the land which has been developed at high costs is planted with poorly husbanded subsistence crops, the return of which is only a few hundred shillings per acre.

The benefits and costs of the three irrigation schemes differ according to the difficulties encountered. Little can be said about Galole, the pilot scheme on the Tana River, except that it should not have been settled but

Table 13. *The economics of rice farms at the Mwea irrigation scheme in 1964*

Irrigated land per farm, in acres[1]	4
Thereof rice in per cent	100
Yields, lbs paddy per acre	4960
Receipts, shs per holding	3733
Home consumption of rice, shs per holding[2]	279
Total, shs per holding	4012
Expenses, shs per holding	550
Water rate, shs per holding	800
Total, shs per holding	1350
Accounting profit, shs per holding	2662
Wages for hired labour, shs per holding[3]	250
Family income, shs per holding	2412
Total days worked per farm[4]	880
Total production per day worked, in shs[4]	4/6
Accounting profit per day worked, in shs[4]	3/0
Days worked by paid labour in per cent of total days[4] worked	44

1 In addition, surrounding unirrigated land is farmed by the tenants or grazed by their cattle.

2 Rice only. The above estimate is probably on the lower side. According to an estimate of the Farm Economic Survey Unit in 1961 and 1962, shs 469 and 486 worth of paddy were retained. However, the estimate being based on the number of sacks issued and not returned is probably unreliable as well.

3 According to a survey made by the Farm Economic Survey Unit in 1963/64, shs 1032 were spent for paid labour. The above estimate is probably on the low side.

4 According to the survey of the Farm Economic Survey Unit. The family income would be correspondingly lower if the percentage of the work done by paid labour is actually as high as estimated by the FESU.

Source: Department of Agriculture: Annual Report, Mwea Irrigation Scheme, 1964 — Farm Economic Survey Unit: Labour Costs and Utilisation in Rice Production on the Mwea-Tebere Irrigation Scheme.

run as an experimental large farming unit. The benefits of Mwea-Tebere have already been significant. In 1964 the scheme marketed 166,000 bags of rice valued at £ 268,000 from land that, prior to development, supported no permanent population. In addition, a certain amount of rice is kept for home consumption and black market sales. The deliveries cover 90 per cent of Kenya's market demands for rice. Approximately 10,000 formerly landless people are being supported by the scheme. In 1964 tenants in the Mwea section received an average income of shs 2,662/- (see Table 13), most of them obtaining some additional income from dry-land farming and cattle grazing on land surrounding the irrigated area. A remarkable feature of Mwea-Tebere, as of all other settlement schemes in East Africa, is the wide differentiation in income between the tenants owing to difference in husbandry and labour efficiency, although the farming conditions are almost exactly the same for everybody.

Total investment for Mwea-Tebere from 1954 to 1962 amounted to £ 881,000 (net), i. e. £ 145 per acre. This figure has to be interpreted with caution. On the one hand, part of the costs must be attributed to the special circumstances of the Emergency. On the other hand, the engineering costs have not been carried by the scheme, but by another Government Department. Considering these two adjustments, it seems appropriate to take the above figure of £ 881,000 as the relevant one. We then arrive at a capital-output ratio of 3.3 : 1 which is rather favourable by international standards.

A comparison of benefits and costs of the Perkerra scheme, on the other hand, gives a most unfavourable result. To start with, there was no remunerative cash crop and year after year substantial annual losses accumulated. Subsequently a suitable cash crop was found in a fungus-resistant variety of onions. In 1962, when onions valued at £ 21,000 were sold, the total production of the scheme hardly exceeded £ 40,000. Total costs, however, had up to 1964 accumulated to £ 416,000. This would correspond to a capital-output ratio of 10 : 1, which is very poor by any standard. Even years after its establishment the scheme is not in a position to cover running costs. The annual subvention amounted to roughly £ 20,000 during the last four years. So far Perkerra has not been an asset but a drain on the national economy, although with proper discipline and different tenants it could probably be profitable or at least cover running costs.

3. Evaluation

With regard to most settlements on new rain-fed land, we can conclude that the objectives were not reached, that the number of families and stock moved out of densely settled areas was negligible, that the schemes failed to introduce a sound system of farming, and that there was little additional marketed production. Most of the planned settlements happened to be in an uneconomic intermediate stage between extensive and intensive types of settlement. In most cases it would have been more economic and less costly just to hand over the land to expanding peasantries as in Chepalungu.

The best that can be said about most of these settlements is that they were highly uneconomic measures of social relief. The poor performance in most cases is, in our opinion, not due to any great number of avoidable mistakes made by the personnel employed. Even the best Settlement Officer and the most efficient settlement agency could not create economically sound schemes on marginal land without a high price cash crop, with a great number of lethargic settlers who had been pushed into the settlement by an administration which wanted to get rid of them, and without sufficient Government authority to enforce sound husbandry. There was nevertheless one great advantage which should not be overlooked: the British agricultural personnel learned the lesson that, if settlement is to succeed, it must be done thoroughly, with heavy additional loans and with the emphasis on cash crops. The personnel concerned obviously tried to apply these lessons — as far as it was in their power — to the re-settlement schemes in the former White Highlands (see V.).

With *irrigation settlement* there is one great success, Mwea-Tebere, one dismal failure, Perkerra, and — largely because of premature settlement — a difficult pilot scheme at Galole. Mwea-Tebere became a success because:

— There has been proper management. The tenants are willing to follow the orders of the Settlement Officers because their income is high and disobedience leads to eviction. High incomes are the consequence of close supervision.
— Land and water are good and could be developed comparatively cheaply. There is easy access to the scheme. The market, Nairobi, is close by.
— A cash crop with high returns per acre is planted. Monoculture facilitates supervision. The high level of yields are attributable largely to the high degree of control exercised on tenants.
— The tenants are Kikuyu and Waembu, who come from densely populated districts with a rapidly progressing semi-commercialised farming community. In most cases they have no other land, no place to go to, and their livelihood depends on their participation in the scheme. The scheme management is therefore in a strong position towards the tenant. The threat of eviction is proving to be a powerful instrument under these conditions.
— The tenants have to pay substantial water rates which cover the running costs and leave a considerable margin for amortisation and interest. The one-channel marketing for rice makes it comparatively easy to collect water rates and to accumulate funds for a rice stabilisation fund.

At Perkerra good land and good water are available and the distance to the railway and the next major market is manageable. The development of a profitable cropping system is certainly much more difficult than at Mwea-Tebere. The size of the latter scheme makes it easier to ensure proper management. However, most of the other features which distinguish Mwea-Tebere are lacking, largely because of the human element involved.

V. Resettlement in the Former Scheduled Areas

Land has always been the main centre of contention and the driving force behind political agitation in Kenya. It was evident that with the coming of Independence a policy of Africanisation of the "White Highlands" had to be formulated. The most important aspect of this policy was the planned resettlement of Africans on land which had been settled and developed by large-scale European farmers. Most of the resettlement has occurred in the mixed farming areas [1]. The plantation and ranching industry has been touched in a few cases only so far.

Table 14.

A. Resettlement programme in the former Scheduled areas

Year	Acreage purchased	Number of Settlers
	'000	
One Million Acre Scheme		
1962/63	220	5197
1963/64	273	9253
1964/65	241	9000
1965/66	101	7050
Total	834	30,500
IBRD/CDC Schemes		
1961/62	29	423
1962/63	27	399
1963/64	49	405
1964/65	51	1800
1965/66	24	1473
Total	180	4500
Grand Total	1014	35,000

B. Resettlement performance in the former Scheduled areas, June 1965

	No. of schemes	Acreage settled	Families
High Density (Million Acre) Schemes	60	592,356	..
Low Density (IBRD/CDC) Schemes	26	144,079	..
Total	86	736,435	23,917
Co-operative farms [1]	12	165,944	1052 [2]
Being farmed by the Departement		213,580	.
Total		1,115,959	.

[1] Mainly ranching associations.
[2] Members.

Source: (1) Ministry of Lands and Settlement, Nairobi 1965, Mimeo. (2) Annual Report, Dep. of Settl. 1964—1965.

[1] A mixed farm is one having more than 20 acres of land in the former Scheduled areas, on which wholly or partly arable cultivation is carried out. The acreage of coffee, tea or other permanent crops on the farm does not exceed 100 acres. In 1962 the mixed farming areas produced 32 per cent of the total large farm production; 24 per cent of the total agricultural exports; 50 per cent of the agricultural exports to Tanzania and Uganda; and employed 111,700 workers with an annual wage bill of £ 4.4 million. They had in addition to the value of squatter cultivation a net disposable income before taxes of approximately £ 15,240,000, and contributed to Government revenue an estimated £ 1,417,000 per annum.

"Settlement Progress Map" (Kenya Highlands)

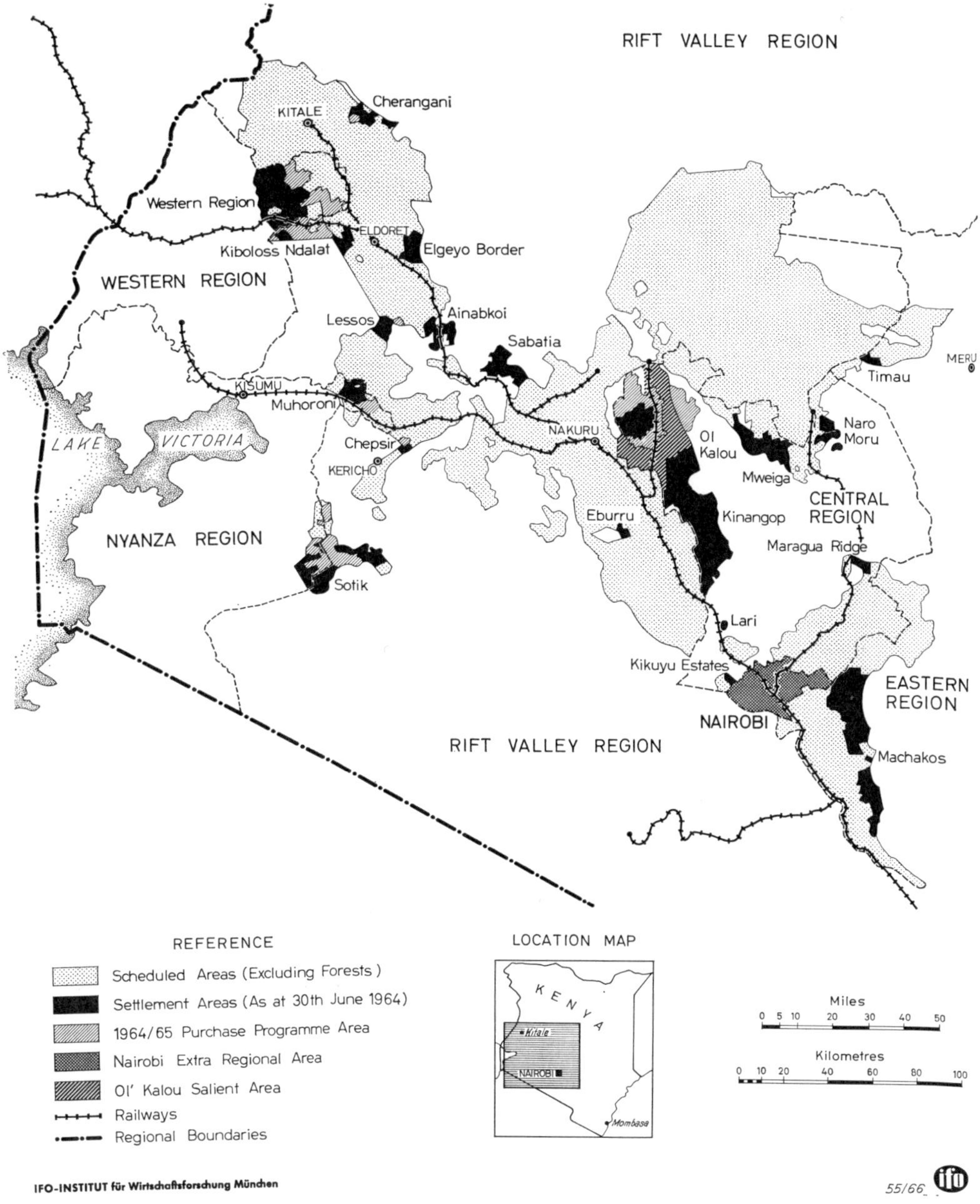

In 1962 the purchase, partition and redistribution was planned of 1,014,000 acres to 35,000 settlers, involving about one seventh of the former "White Highlands" (see Table 14.). By June 1965, 1,180,000 acres had actually been purchased for resettlement and 716,342 acres had already been distributed, or were in the process of distribution, to 25,746 settlers. It is anticipated that the final figure of 34,000 families settled will have been reached early in 1966. Taking the Ministry of Agriculture's estimate that 70 per cent of the former Scheduled areas are, in the main, suitable only for grazing, and, assuming that most of the settlement is taking place on land of high potential, this means that so far (March 1965) roughly half of the high potential land in the ex-Scheduled Areas has been handed over to the African settlers.

1. Aims, Institutions and Methods

a) Aims

The resettlement schemes are inspired by socio-political rather than economic motives. Their main aim is to facilitate the rapid and orderly transfer of land from European to African ownership. However, the actual planning is much more ambitious. Resettlement is expected to combine almost impossible objectives: to increase production and productivity to such a degree that the new settler is (1) much better off than before and (2) in a position to pay back (a) the development loan and (b) two-thirds of the purchase price given to the outgoing British farmer, including a 6.5 per cent interest charge. Furthermore, settlement is expected

— to increase the value of total production from approximately £ 4—5 per acre to an overall budget target of £ 8,3 per acre. See Table 15. The idea is that the large scale farmers, shying away from the difficulties connected with a large labour force, preferred labour-extensive enterprises, i.e. beef-cattle, sheep, wheat etc. The transfer of people out of the densely populated 'Reserve' into settlement areas is increasing the available labour force per 100 acres. Furthermore, African settlers are likely to have access to cheaper non-family labour than had the large-scale farmers whose labour was organised. Both these factors should contribute to an intensification of land use which, it is hoped, will more than offset the lower technical efficiency of the new settlers;
— to relieve the land hunger in the former Reserves by enabling farmers there to buy land from those who are going into settlement areas;
— to improve the unemployment situation, not explicitly by the settling of urban unemployed but by improving the employment situation generally;
— to provide farms for those Africans who were tenants in the former Reserves and have become dispossessed by the process of land consolidation;
— to replace the racial dualism between African smallholdings in the Reserves and large-scale farms in the Scheduled areas by the transfer of farms of widely different sizes to African owners;
— to introduce better methods into a large segment of the rural African population by the supervision of production through the settlement staff;

— to assist in creating a reasonable market for land in Kenya in spite of the precarious socio-political situation for the white minority. Most European farmers are anxious to sell. This would inevitably have led to a disastrous drop in land prices and consequently to chaos in the large farm economy if settlement had not appeared as a major buyer of land who did not intend to exploit the anxiety of the farmers. Thus one of the objectives of settlement was to stabilize the remaining large farm economy.

The provision of substantial loans to the settlers is considered to have been of key importance in attaining these objectives. Resettlement is not just the redistribution to smallholders of the land bought from the large farmers. The basic idea is that the injection of large loans places the settlers in a position to develop their holdings fully in order to pay back the development loan itself and two-thirds of the purchase price.

b) Institutions [1]

In June 1963, with the coming of internal self-government, the African Land Development and Settlement Board (Aldev) was replaced by the Central Land Board, a corporate body deriving its finances from the Government, whose functions are to select, value and purchase agricultural lands for settlement in the former Scheduled Areas and to convey titles to those Africans who are settlers. After the purchase, the land is handed over to the Department of Settlement within the Ministry of Lands and Settlement. Simultaneously with the constitution of the Central Land Board, a body of trustees, known as the Settlement Fund Trustees, consisting of the Ministers for Finance, for Agriculture and for Lands and Settlement, was created to operate an Agricultural Settlement Fund for the financing and management of approved schemes. The Department of Settlement within the Ministry of Lands and Settlement takes over the farms and has to execute the actual settlement and related operations.

The partition and distribution of the land takes three distinct forms:

— The high density Smallholder Schemes for Africans with limited capital and agricultural knowledge. Each holding has to provide for subsistence, financial obligations and a minimum net income of £ 25—70 per annum. The high density schemes are financed by British and some German money. Up to mid-1965 26,222 settlers have been established on 556,110 acres. From the point of view of both the total size of the operation and the numbers involved, this is by far the most important group and is referred to as the "Million Acre Scheme".

— The low density Smallholder Schemes for experienced farmers with some capital. Each holding has to provide for subsistence, financial obligations and a minimum net return of £ 100 per annum. Most of the low density settlements are financed by funds from IBRD and CDC. Up to early 1965, 4,088 farmers — significantly more than originally planned — had been established on 140,098 acres (See Table 14).

[1] See in particular: C. P. R. Nottidge and J. R. Goldsack: The Million-Acre Settlement Scheme 1962—1966, Dep. of Settlement, Nairobi 1965.

— The yeoman schemes for experienced farmers with substantial capital. Each holding has to be sufficiently large to provide the settler and his family with subsistence, the means of meeting his financial obligations and a minimum cash income of £ 250 per annum. Owing to the heavy capital demand in relation to the number of families concerned, only a few settlements of this type have been established. By direction of the Cabinet, a new policy was started in 1964 whereby the better class houses on large scale farms had a 100 acre holding planned around them, regardless of the size of the plots in the remainder of the scheme. This was done so that the house and 100 acre holdings could be sold to a leader of the community such as a member of the Central Assembly or a Senator (Annual Report, Department of Settlement, 1963/64 p. 5). The schemes are financed from British funds.

Table 15. *Budgeted production performance of matured resettlement*[1]. *Involving 779,375 acres and 33,317 settlers*

Pyrethrum	£ 435,000	
Wheat	£ 122,000	
Maize	£ 742,000	
Tea	£ 325,000	
Sugar Cane	£ 1,242,000	
Sheep	£ 35,000	
Dairy Cattle	£ 2,563,000	
Subsistence	£ 666,000	
Total	£ 6,467,000	— £ 8.3/acre.

Pre-settlement total production value[2]: £ 4—5 per acre
Actual Settlement total production value[3]: £ 5—6 per acre

[1] It is assumed that the settlement will normally reach maturity within four years after the take-over, except where tree crops like tea have been established.
[2] Rough estimate based on data supplied by the Farm Economic Survey Unit. (See Table 19).
[3] Four schemes in their first full year. Compare Table 19.
Source: Department of Settlement. Nairobi 1965, Mimeo.

The Schemes are planned into units of roughly 10,000 acres for high density areas and 5,000 acres for IBRD/CDC, i.e. low density areas. Each scheme is staffed by a Settlement Officer — frequently a former European farmer — and auxiliary administrative, agricultural and veterinary staff. Between seven and ten Settlement Schemes are grouped together under the command of a Senior Settlement Officer who in turn is answerable through a Settlement Area Controller to the Director of Settlement. To a large degree the trained administrative and technical staff of the Department of Settlement has been seconded from the Ministry of Agriculture. The shortage of personnel has led to the employment of American Peace Corps volunteers in resettlement schemes. They are doing a most useful work as assistants to Settlement Officers or even as Settlement Officers themselves.

The settlers have to join a Producer Marketing Co-operative, which arranges for the local collection and bulking of the produce and for its sale to established agencies. Certain commodities like wheat, pyrethrum and milk have to be sold through the co-operatives. As soon as it is feasible, the Settlement staff is expected to hand over communal services such as dipping, artificial insemination, tractor ploughing etc. to the co-operatives which are also actively encouraged to handle the repayment of dues. The first co-operative in the Settlement Areas was registered in 1962. By the end of 1965 there were fifty-five co-operatives with a total membership of 10,700.

Thus resettlement is an arrangement between three partners who are directly concerned and responsible. (1) The individual settler, with conditional freehold title to his plot, who is subject to membership in (2) a co-operative, and to supervision by (3) the settlement staff, who also supervise the performance of the co-operative.

Although there has been considerable discussion about co-operative land use in resettlement areas, no action along this line had been taken up till 1964, with the exception of two large co-operative ranching schemes, the organisation of which corresponds much more to a company than to a co-operative in the classical sense of the word. In 1965, however, several co-operative wheat-sheep farms on soils unsuited for smallholder settlement had been started.

c) Methods

The criteria for choosing the settlement areas have been political and social rather than agricultural or economic. The following considerations seem to have been decisive:

- The land should be contiguous to or at least not far from the tribal land unit. This has two obvious advantages. (1) It proves to the tribes concerned that the very land they were claiming has been given to them. (2) Settlement thus does not destroy Lord DELAMERE's idea of a concrete belt of large farms in the Highlands. There is still a contiguous belt of large farms, ranches and forest reserves, stretching almost from Nairobi to the Uganda border.
- The need to eliminate islands and salients of European farming surrounded by or jutting into areas of African settlement.
- The land should be close to former Reserves with a high population pressure.
- All of the major tribes bordering on the former White Highlands had to be provided with some settlement land. The extent of the land obtained for different tribes was determined by a sub-committee of the council of Ministers and varied from 400,000 acres for the Kikuyu down to some five or six thousand acres for the West Pokot.
- The land should be of a high potential and largely underdeveloped.

The purchase of the land was, and is, on a "willing buyer — willing seller" basis [1]. Cases of farmers being unwilling to sell have been extremely rare. (A number of European farmers, however, after the sale of their farms

[1] So far experience has shown that a large farm surrounded by smallholders is subject to so many thefts of stock, crops and fencing-wire that economic production cannot be maintained.

have bought farms elsewhere in Kenya, thus creating something of a general land market.) The valuation of farms has not been uniform. In some cases they were bought on the basis of 1959 land values, in others they were valued on the basis of eight times the average annual profit, working on a 12.5 per cent return on the capital invested in land, buildings, water supplies, roads etc [1]. The outgoing farmer disposes of the movable assets. The price for grade cattle is being kept very high indeed, largely because the Settlement Authorities are buying grade stock in order to equip in-going settlers. The price for land and immovable assets was calculated at an overall average of £ 9/acre, or roughly £ 9 million for the one million acres concerned. The experience gained up to 1965 has shown that the actual average price paid is around £ 11—12 per acre, i.e. £ 11—12 million for the current programme.

Once the land has been purchased, it is a matter for the Department of Settlement, in close liaison with the Provincial Agricultural Committees and the Provincial political body, to decide whether a High or Low Density Scheme is appropriate in the particular area. Roads, soil conservation works and boundary lines of each holding are established before the settlers move in. Until these works are constructed the Department of Settlement maintains the land in production.

The responsibility for the selection of settlers rests with the Provincial political body (Provincial Advisory Committee), which establishes a local tribal committee for this purpose. Applicants who have been approved draw their plot numbers from a hat. The criteria for the selection of applicants differ from place to place. In some cases there were hardly any yardsticks, in others preference was given to the landless and unemployed, and again in others to applicants of wealth and repute. Civil servants have no influence whatsoever in the selection of settlers for High Density Schemes. Only in the IBRD/CDC Schemes and in the Yeoman Schemes applicants were required to have some agricultural experience and sufficient working capital.

A budget is prepared for all plots in accordance with local circumstances. The functions of the budget are (1) to indicate the plot size required to meet the given income target, (2) to suggest a basic loan structure for the individual farmer and (3) to provide a framework upon which the incoming settler can base his farm programme.

---

[1] The only lender who supplies money for the land purchase is Britain. Funds from the IBRD, CDC and the Federal Republic of Germany are used for development purposes only. Britain, in paying 8-times the average annual profit to outgoing European farmers, including non-British citizens — there have been a great number of South Africans and Scandinavian farmers in the former Scheduled areas —, is certainly granting one of the most generous compensations for political losses of landownership known in economic history.

The settler is expected to meet the conveyance, registration and stamp duty charges. In addition, he has to pay a membership fee to the co-operative in his scheme. In Low Density and Yeoman Schemes a down-payment varying between 1,000 and 5,000 shs has to be made, which is deposited with the Department of Settlement as working capital. Thus a settler in a High Density Scheme has to pay approximately £ 6 in order to start, while £ 100 and £ 500 respectively are required in a Low Density or Yeoman Scheme (see Table 16).

The balance between the settlers' contribution and the total costs — not counting costs of overheads, staff and settlement preparation — is shouldered by

— a subsidy, in the form of a grant from the British Government, which amounts to one third of the purchase price of the land and immovable assets;
— a purchase loan, the principal of which has to be repaid over thirty years at 6.5 per cent interest;
— a development loan (for grade cattle, fencing, tea stumps, tractor cultivation, fertilizer, seeds and — the only household item — iron roofs) which the settler can obtain on application (Not all, but most of them apply). This is repayable over ten years at 6.5 per cent interest. The purpose of the development loan is to get the holding operating at an economic level. In disbursing development loans no cash payments are made to settlers.

Table 16. *Estimated finances involved, per holding, according to type of resettlement scheme (excluding administrative costs)*

Type of Scheme	Costs of Land to Settler	Purchase by Subsidy	Development Loan	Settler's own Capital Requirement	Total costs
	£	£	£	£	£
Yeoman	440	220	1,160	500	2,320
Low Density	145	75	430	100	750
High Density	95	45	100	6	246

Source: ETHERINGTON, D. M.: *Land Resettlement in Kenya: Policy and Practice* (adopted from a talk by STORRAR, A. "Resettlement in Kenya").

All loans have to be paid in equal half-yearly instalments without any moratorium [1]. Altogether the total costs of one settlement amount, according to STORRAR's estimates, to £ 246 in High Density Schemes, £ 750 in Low Density Schemes and £ 2,320 in Yeoman Schemes. The proportion of the total costs provided by the settler declines from nearly 22 per cent for the Yeoman, to about 13 per cent for the Low Density participant, to less than 3 per cent for a settler in a High Density Scheme.

Table 17 shows the budget of a 29-acre holding in a Low Density Scheme. The establishment of the holding costs Shs: 14,900. Of this, 5,000 shillings is a sub-

[1] The systems of loan repayments differed in the earlier schemes.

sidy called Settlement Charge. A loan for the purchase of the land amounting to Shs: 4,500 has to be repaid over 30 years at 6.5 per cent interest, in annual instalments of shs 344. The Development Loan of shs 5,400 has to be repaid over

Table 17. *The budget of a low density scheme holding, Lietego number 94*

Size of the holding: 29 acres.

Target income	— £ 100 plus subsistence
Settlement charge	— 5,000/—
Land loan	— 4,500/— (30 years @ 6 ½%), annual repayment 344/—
Development loan	— 5,400/— (10 years @ 6 ½%), annual repayment 742/—

Loan Schedule	shs	Qualifying capital	shs
Livestock	3,750	Stamp duty	100
Buildings	600	Legal fees	40
Fencing & tools	600	10% deposit of land	500
Establishment of crops	450	One cow valued at	500
	5,400/—		1,140/—

Farm plan at maturity (3rd year)	Livestock inventory	L. U. S.
0.5 acre homestead	8 dairy cows	8
2.0 acre subsistence	2 heifers	2
3.0 acre maize	2 yearlings & 2 calves	1
1.0 acre ley, undersown		11
22.5 acres grazing		
29.0		= 2.2 acres per L. U.

Income per annum	shs
3 acres maize, 12 bags p. a. @ 29/— per bag	1,044
8 cows, 130 lbs butterfat @ 2/50 per lb	2,600
2 cull cows @ 300/— each	600
4 surplus calves (2 kept) @ 30/— each	120
	4,364/—

Expenses, per annum	
3 acres maize @ 143/— per acre	429
8 cows @ 70/— per annum	560
2 heifers @ 70/—	140
2 yearlings @ 35/—	70
1 acre ley, undersown @ 60/— p. a.	60
	1,259/—

Total gross income	4,364
Total expenses	1,259
Total loan repayments	1,086
Cash margin to farmer	2,019/— (£ 101)

10 years in instalments amounting to shs 742 per annum. Thus in the first 10 years shs 1,086 have to be paid per annum by the settler and thereafter shs 344. In addition, the settler, in order to qualify, is required to pay a down-payment of shs 1,140, of which shs 500 is for the land and shs 500 is used for the purchase of

one grade cow. Most of the loan (shs 3,750) is used for the purchase of additional grade cattle. Three years after the settlement, the holding is expected to comprise 2 acres of subsistence farming, 3 acres of maize for sale, 1 acre of ley and 22.5 acres grazing; 8 dairy cows and 2 heifers. The income then is estimated at shs 4,364 and the expenses, without payments to wage labour, at shs 1,259, which leaves the settler, after the loan repayments have been deducted, with a cash margin of shs 2,019. In addition, he has his subsistence food.

The settler obtains a freehold title subject to certain conditions. The land shall not be subdivided, charged, let, leased, or transferred without the prior consent of the Central Land Board. The allottee shall, within six months of the date upon which he takes possession of the land, cultivate at least one acre of arable land, erect a dwelling house, build a fence or plant a hedge around the perimeter of his land. The provision of loans is tied to the condition that the borrower farms his land in accordance with approved methods of good husbandry to the satisfaction of the Settlement Fund Trustees and that he pays all rates, charges and instalments. Cattle bought on a loan may not be sold or slaughtered and only cattle checked and approved by the Veterinary Officer are allowed on a scheme. The settler has to dip his cattle weekly. On most schemes he may not keep scrub bulls, but has to have his cows served by artificial insemination. According to a provision in the settlement contract settlers who do not comply with these rules and who do not pay their loan instalments are liable for eviction. Hovewer, no other rules concerning good husbandry have been promulgated, except the few ones mentioned before. The emphasis is laid on persuasion and advice, not on supervision and enforcement.

2. Difficulties, Benefits and Costs

a) Difficulties

The settlement of 33,000 families on more than a million acres within four years is a bold venture, particularly during the transitional period from a colonial to an independent administration. A programme of such scope is bound to run into expected and unexpected problems. The following difficulties seem to be of particular importance:

α) The Crash Programme

Many of the present shortcomings of settlement can be traced back to the speed with which everything had to take place. Resettlement started rather late. Political pressure was building up in 1963 and 1964. In order to avoid chaotic landgrabbing, a "crash programme" involving huge purchase of land had to be carried out at extremely short notice. 3,000 families were settled within 21 days on the Kinangop, a high plateau claimed by the Kikuyu on which organised settlement had started in 1962. The settlement staff was hastily recruited and appointments were made with little checking of the appointee's qualifications. Serveral schemes, not only those on the Kinangop, started before the Agricultural and Veterinary staff could be seconded. In some High Density Schemes, the land which was doled out is now proving to be utterly unsuitable for smallholders: it has heavy clay soils; it is poorly drained; the area is too cold and even frost-prone and not at all adaptable to the crops which fit into a smallholding of a few acres.

β) Tribal Rigidities

Very little of the original idea that the settlers should be the cream of the smallholders in the Reserve could be realized. The selection was based in the first instance on tribe, which is certainly a realistic approach considering the tribal frictions and rivalries still existing. The settlement schemes can actually be regarded as tribal overspill areas. Expedient as this might have been, it certainly did not promote optimum land use: some land had to be given to tribes who did not need it; backward tribes had to be settled on land of high potential, although it would not be well developed by them; in a number of cases, stands of coffee could not be allotted to individuals with experience in coffee, but had to be handed over to uninterested settlers of the local tribe who had only a pastoral background.

Farm workers who had previously been employed by the large-scale European farmers and who did not belong to the tribe in question were frequently driven away or, at best, accepted by the settlers as squatters or as badly paid labourers. Settlement has been particularly hard for the Kikuyu, who formed a large percentage of the labour force on large farms. It is true that the Kikuyu obtained the greatest chunk of settlement land, but much, in fact most of it, is basically unsuitable for smallholders. It is open to question whether a comparison of the economic loss to the Kikuyu farm labourers, with the economic gain to the Kikuyu settlers, would reveal an economic gain for this tribe as a whole — the tribe which carried most of the burden of the fight for the land. On the other hand, the members of the Kalenjin tribes clearly profited. They were in a position to claim comparatively large stretches of high potential land. Consequently, they obtained larger farms in Low Density Schemes and larger loans as well; and this in spite of the fact that the Kalenjin — with a substantial number of notable exceptions — are far less commercially minded and industrious than are most members of the Kikuyu or Kisii tribes.

γ) Arbitrary Selection of Settlers

Another set of difficulties was connected with the absence of proper selection of settlers. In most High Density Schemes little or no regard was paid to the agricultural qualifications of the applicant. Preference was given to people who had friends in the committees or were said to be poor or landless and who, more often than not, had failed somewhere else or were unemployable. The number of High Density holdings where the settler is now looking for outside employment, while his wife is scratching the surface of the plot trying to grow some subsistence crops, is probably rather large.

Settlers in Low Density Schemes frequently have divided interests: a farm in the former Reserve, a shop in town, a position with the administration, etc. There are — so the rumour goes — settlers with several plots. In some schemes not more than 50 per cent of the settlers are assumed to have the holding as their only means of livelihood. The rest might be considered part-time settlers, employing relatives or labourers to work on their settlement enterprise. This arrangement need not lead to poor results. A significant proportion of the better farms in Kenya are owned by absentee ownners, who invest their non-farm savings in their holdings. However, many of the settlers with divided interests do not have enough capital to run their farms. Thus it comes about that some of the plotholders, who are outstanding farmers in the former Reserve, prove to be poor settlers and consider their plot as a kind of land insurance policy for the family.

δ) Shortcomings in Animal Husbandry

It can definitely be said that no other item is as important for the success or failure of the settlement schemes as the successful husbandry of dairy cattle. There is general interest in commercial milk production although most of the settlers have had no prior experience with grade cattle. The settlement staff have succeeded so far in keeping losses through disease within reasonable limits. A remarkable acreage has been put into fodder-crops by people to whom the principle of feeding animals was unknown before. Hopeful as these indications might be, there remain two highly important problems to be solved:

— In order to control breeding disease, bulls have been forbidden in almost all schemes and artificial insemination has been introduced. On a few schemes A.I. is proving a success. But in many other cases the situation is quite the opposite. Discussions on the spot with a great number of settlers and Settlement Officers indicate that the actual calving interval in the settlement schemes owing to poor feeding, insufficient attention on the part of the settler and shortcomings of artificial insemination is likely to average 18 months. It is obvious that this shortcoming alone — if it is allowed to continue — is incompatible with an economic milk production, with the repayment of loans taken for the purchase of cows, and thus with successful settlement. One of the main problems being the shortage of heifers to stock settlement fully with grade cows. The prices for heifers have already doubled in the last three years, largely because of the long calving interval coinciding with a high demand on the part of the African farmers.

— In addition, it has to be pointed out that a high percentage of the few heifers which are born are not properly reared. Their demand for milk competes with the desire for cash and for home consumption, and the calves are thus not obtaining what they need to develop properly.

These two shortcomings are certainly a great danger to the dairy herd of Kenya, a substantial part of which — according to the plans — will gradually be transferred into the settlement schemes.

ε) *Problems of Land Use*

The idea that certain income targets should be reached has led to the establishment of holdings which are frequently larger than is usual in the former Reserves. Consequently, the plots are often too big in relation to the work performance of the settler and his family. The essential shortcoming is not one of farm size but lack of proper land use. The first reaction of the settler is to get his subsistence crops and his house established as quickly as possible. Regardless of soil and climate, maize is planted everywhere. However, most of the settlement areas are at high altitudes where maize yields are low and where it takes ten, twelve or even fourteen months for the maize to mature. It is obvious that the effort devoted to maize is highly uneconomic because other profitable activities are being neglected. One of the greatest disappointments in 1964 was the size of the pyrethrum production. The settlement areas produced only 44 per cent of their quota as compared with the whole country's average of 70 per cent.

Much more intricate are the problems posed by those activities which do not fit easily into smallholdings; a case in point is the production of wheat. Hesitant efforts to divide the land suitable for wheat cultivation into large blocks where the individuals would have their plots met with the opposition of the settlers. The idea of enclosed individual holdings is dear to them. Hence wheat production has to be carried out in a large number of small fields. It was intended that the ploughing, planting, fertilizing and combining should be organised by the co-operatives, but this proved to be an unrealistic proposition. At best the Settlement Officers manage the wheat production, while the settler sits at the edge of the field watching the contractors do the work.

In some cases, however, the settler has tried growing wheat on his initiative with quite disastrous financial results. The gross returns, minus the costs to the contractors, go to the settler whose position is one of a rentier receiving unearned income while those who manage the production, the Settlement staff, are paid by the Government.

A similar situation is likely to arise in sugar-settlements, as carried out in Muhoroni. The sugar production is mainly organised — and rightly so — on a plantation basis. The settlers receive plots in larger field units which undergo planting and cutting. However, half of the manual work has in any case to be done by plantation labour. The only work that can be allotted to the settler is weeding, fertilizing and trash clearing. But the settler will receive all the profits although he did not contribute the capital — which was given on loan — and only half of the labour. He is thus obviously profiting from the efforts of others. In our opinion, in the case of sugar production the idea of combining settlements with plantation crops has definitely been carried too far.

The same is likely to become true for sisal in other settlement schemes. Hedge sisal is certainly an interesting sideline for any smallholder. However, the planting of small compact sisal fields on a great number of widely spread smallholdings means such high costs for cutting, transport and decortication that production can be economic only if the price for sisal fibre is very high indeed.

ζ) Increased Erosion

In most of the settlement areas vast acreages of extensive pasture have been broken up and put under arable farming. This means intensified land use and must basically be welcomed, although yields per acre are rather low. It should be realised, however, that this also implies the mobilisation of the great capital in soil fertility which has been lying dormant in leys, pastures and bush and which will be wasted by soil mining if the usual methods of African soil husbandry are applied. There is a tendency to open up land which is generally considered too steep for cultivation, and while there is some increase in production because these slopes are usually rather fertile, the increased silting up of rivers is already causing concern after the very first year of settlement.

η) Hoe, Ox-plough or Tractor?

Most of the holdings in High Density Schemes are too large for the hoe and too small for a tractor or even a pair of oxen, while most of the holdings in Low Density Schemes are too large for an ox-plough and too small for a tractor. The idea that the co-operatives would organise the ploughing of the land is probably unrealistic. Most of the cultivation up to now has been organised by the Settlement Officer — frequently a former British farmer — who hires contractors — also former British farmers. It is open to doubt whether such a set-up will be accepted for any length of time. In some locations successful settlers have already established themselves as small scale contractors, doing the ploughing for their neighbours. Generally, however, a satisfactory answer to the problem of timely cultivation has not yet been found. The weakness really stems from settling smallholders on land best suited to the cultivation of small grains on large acreages.

ϑ) Non-Fulfilment of the Budget Targets

All the above difficulties obviously affect production and productivity. According to data collected by the Farm Economic Survey Unit and interpreted by R. H. Clough, the value of marketed sales in the first year of settlement has reached only 60—70 per cent of the target, while expenses have been higher than expected, thus leaving a balance which is hardly sufficient to repay the loans. It has, of course, to be considered that the first year of settlement is not likely to bring full production. However, it seems that a substantial discrepancy between the performance that can be expected and the targets laid down in the budget is likely to remain. The budgets have certainly been cautious as to yields per acre and per animal and as to prices, but they apparently aimed too high with regard to:

— the scope of the enterprises a family can manage without the help of paid labour. The budgets overestimated the acres of maize, the number of cows, the hours of labour a settler is willing to put in, and the labour efficiency within the working time;
— the time in which a plot can be brought to maturity. Settlement planning is usually based on the assumption that the settler is straightforward and aspires to farm according to the budget. Experience shows, however, that most of the settlers, particularly those in High Density Schemes and with little capital, seem to progress in distinct phases. Phase (1) is the establishment of the house, the fencing of the holding and the planting of subsistence crops. Phase (2) is characterised by increased planting of cash crops, in particular of pyrethrum and potatoes, which seem to be much more rewarding than any other activity. Finally in phase (3) more attention is given to the animal husbandry side, which in any case can only slowly be expanded because of the exorbitant prices for heifers. A holding can be defined as having reached maturity when the value of its production differs in detail only with the budget target. Settlement planning assumes that most schemes will reach maturity by their fourth year, except in cases where particular circumstances prevail such as, for instance, holdings with newly established tea. This assumption is probably too optimistic;
— the availability of grade stock. The prices of grade stock have gone up because of the high demand of settlement and of the Reserves and because of reduced production owing to fewer large farms and long calving intervals of grade cattle owned by Africans.

Hand in hand with the non-fulfilment of budget targets are the difficulties of loan repayment. The half-yearly instalments are a heavy charge in any case, and the burden is likely to be unbearable for those who are not producing according to the plan. This is particularly true in the first year: there is no moratorium, and the first loan instalment is already due half a year after the take-over [1]. This fact is certainly not clear to all settlers. Many of them do not realize the implication of a loan. If they do not improve it will be impossible for them to pay their instalments out of their settlement enterprise.

ι) Overburdened Co-operatives

The word "co-operative" seems to be considered a magic cure for most of the economic shortcomings of the present African scene. The role att-

[1] The fact that the first instalment is due within the first six months after the settler gets onto the land is the cause of a great deal of complaint. It is obvious that the settler will not be in a position to repay the first instalment out of the returns of his holding. The first instalment must therefore actually be considered as part of the purchase price. Settlement planning is apparently based on the assumption — which is probably correct — that most of the settlers are in a position to pay the first instalment out of savings or through the help of family relations.

ached to co-operatives in settlement schemes is a case in point. They are not only expected to sell the produce and to buy agricultural supplies, but in addition to buy and run the dips, to organize artificial insemination, to hire contractors or to run tractors themselves, and to organize loan repayments. It is obvious to any observer of the present situation in East Africa that settlement planners demand far more of the newly established co-operatives than they can possibly give (actually much of the produce is not sold through the co-operatives. This is particularly true of maize). A settlement scheme involving 500 and more producers is big business which cannot be run by a committee. For the marketing side alone a well paid and experienced manager is required, having far-reaching freedom in making decisions and receiving substantial monetary incentives. As regards the other functions mentioned above, with the conflicting interests of members, it may well be asked whether co-operatives are an appropriate institution in the first place. Perhaps it would have been better if the Department for Settlement would have established a Settlement Officer as a benevolent landlord, who — like the landlord in most Asian countries — is handling all those activities which cannot properly be performed by the individual settler.

Table 18. *The performance of the settlers in some resettlement schemes (first year of full settlement, shillings per holding)*

Number of Farms	High Density Schemes				Low Density Schemes			
	Ndalat Results 30		Lugari Results 21		Ainabkoi Results 22		Lessos Results 27	
	Budget Sh	Actual Sh	Budget Sh	Actual Sh	Budget Sh	Actual Sh	Budget Sh	Actual Sh
Output net of home consumption	2077	1259	820	585	6350	3777	3550	2258
Expenses other than loan repayment[1]	847	625	220	407	3602	2179	1610	1127
Loan repayment instalments[2]	514	443	227	188	624	726	566	708
Net profit not taking account of home consumed output	714	191	373	—10 (Loss)	1606	872	1374	423

[1] Including depreciation not in the budget.

[2] The figures about loans are not yet conclusive. The policy has been changed. Thus actually more had to be paid than proposed in the budget in the cases of Anabkoi and Lessos. The payments in Ndalat and Lugari concern the interest element only.

Source: Clough, R. H.: Some economic aspects of land settlement in Kenya. Egerton College 1965, Mimeo.

κ) Insufficient Authority and Supervision

There is the provision that settlers have to follow certain rules and regulations ensuring good husbandry and thus secure the economic success of the schemes. But in actual practice the settlers do pretty much as they like. The settlement personnel are hardly in a position to insist on the growing of certain crops on certain acreages and to enforce rules and regulations. The staff does not have the political background necessary for the enforcement of proper husbandry. Many of the Settlement Officers have been overstrained by the weight, the difficulties and the frustration of their job. It is a major achievement that regular dipping has been carried out normally and that the invasion of illegal cattle has in most schemes been prevented. Thus the personnel is unable to take advantage of the opportunity for intensive and successful extension offered by the institutional framework of settlement.

The diversity between good and poor settlers within each scheme is extremely great. The danger exists that the poor ones are setting the standards. Since the settlement staff has hardly enough authority to clamp down on the worst defaulters, most of the settlers farm as they like, with little regard to what they have been told. There is a danger that in a few years' time some of the settlement areas will look like the poorer parts of the former Reserves and that consequently loans will not be serviced. This is particularly true of some of the Low Density Schemes.

b) Benefits

Impressive as the list of difficulties and shortcomings might be, it should not be overlooked that a number of substantial benefits of the resettlement programme are already recognizable. These have to be considered in order to arrive at a realistic evaluation. Until the settlement has been completed, it is certainly impossible to pinpoint benefits and costs. However, three years after the beginning of settlement some conclusions about the nature and the importance of benefits and costs can undoubtedly be made.

α) Political Benefits

Of paramount importance is the immediate beneficial effect which settlement has had on the political scene. In some areas there was an imminent danger of a land grab. The "crash programme" mentioned above successfully prevented chaos and all the economic losses which would probably have gone along with it. The present Government is now in a position to show that a substantial step in the Africansation of the former White Highlands has been made.

β) Social Benefits

Furthermore, the settlements have greatly diminished the dualistic character of Kenya's society. After the completion of the current schemes,

35,000 families will have distinctly defined titles to pieces of land the potential of which — if properly developed — will allow a standard of living very much higher than the present average in the former Reserve. In addition land settlement has brought a greater degree of social stratification into the rural African society, which might be conducive to the speeding up of social change. Farms in Low Density Schemes in particular give much scope for the development of entrepreneurial talents, and considering the efforts already made, it can be assumed that the challenge will be taken up by some of the settlers. In later years this might perhaps be considered one of the most important of all the benefits of settlement.

A great number of people who had no previous experience whatsoever have been introduced to modern farming techniques: the husbandry of grade cattle, application of fertiliser, use of sprays, growing of fodder crops, etc. An institutional framework has been set up — with the settlers, the co-operatives, and the settlement staff as the parties concerned — which can be used not only to establish firmly those innovations which have already been introduced, but also to bring in new ones such as hybrid maize or the clover-grass mixture for leys.

Although the impact of settlement on the employment situation in Kenya has hardly been decisive, it is nevertheless important. The experience in High Density Schemes thus far indicates that the number of new landowners is 25 per cent greater than that of former labourers. It is estimated that approximately another 25 per cent are taken up in employment by the settlers. This adds up to an increase of employment of roughly 50 per cent, which — as the Department of Settlement hopes — might increase to 100 per cent when the schemes have reached maturity [1]. Low Density Schemes generally absorb the same number of plot-holders as there were labourers on the farm before the take-over. In addition, each plot-holder usually employs one outside labourer. This adds up to a 100 per cent increase in the number of people absorbed.

γ) Economic Benefits

The available information indicates that settlement will increase the value of total production per acre. Table 19 shows the results of some schemes in their first full year after the take-over. According to these estimates, which might not be strictly comparable, the average total of output is 15 to 90 per cent higher than that of the European farmer. Even the sales, by and large, surpass the European farmer's performance. The estimates of Table 19 have of course to be interpreted with caution. Lugari was a poorly developed scheme in 1963, but Ainabkoi is among the very best and Lessos and Ndalat are certainly above the average schemes. Neverthe-

[1] L. H. Brown, on the other hand, who analysed a sample of more mature settlements, comes to the conclusion that the additional employment does not exceed 10 per cent.

less, the estimates are an indication that production and marketed sales — in spite of the obvious shortcomings in animal husbandry and land use — might not be so bad after all. If the budget targets are reached within the next four or five years — which is, however, unlikely — the value of total production and marketed sales would almost double. In any case, we have to attribute some increase in production and sales and all the benefits accruing from this to settlement.

Table 19. *Comparison of the economic performance of farms before and after resettlement*

Scheme	Average Farm Size	Value of total Production Before[2] Settlement	After[1] Settlement	Value of Sales Before[2] Settlement	After[1] Settlement
	Acres	shs	shs	shs	shs
Ainabkoi, Low Density	40	100	115	92	102
Keben, Low Density	24	84	124	71	95
Ndalat, High Density	18	84	110	179	57
Mweiga, High Density	19	111	210	105	153

[1] First full year of settlement.
[2] Rough estimates, which may be on the low side.

Source: MacArthur, J. D.: Estimates of the Farm Economic Survey Unit. Mimeo, 1965.

Much of the "one million acre scheme" concerned undeveloped land or extensive pasture. Settlement brought more and better roads, additional water supplies and fencing. Large acreages of bush have been cleared by the settlers and put under arable crops. Plantation crops like tea und sugar cane have been established at the settlement's expense. Althought the productivity per man and per animal is not as high as it used to be under the former large farmer, there is much more development activity going on in most settlement areas simply because additional capital is being poured in and the labour supply is 50 or 100 per cent greater than before. The main aim of settlement, viz. the change in the man : land ratio, is certainly having effect.

Settlement has obviously contributed to the continued functioning of the remaining large farm economy. The fact that there are potential buyers for land has kept the land prices at a remarkably high level considering the political situation. The large farm economy benefits particularly from high prices for grade cattle. There are more run-down farms in the Highlands than there used to be, but as a general rule the large farm economy has not broken down. The evaluation of farms for settlement purchases according to the annual profit is a compelling reason for most people to keep their enterprise in good order. There are even cases where Europeans have bought derelict farms and developed their production for the sole purpose of selling them, with a substantial bonus for development, to the Settlement Fund Trustees. On the other hand, settlement has also relieved some of the land pressure in the Reserves. Land prices there have dropped significantly. It might be assumed that some part of the land in the former Reserves has been bought by those who want to expand their commercial activities.

In addition, attention needs to be drawn to a change in farm inputs, which certainly corresponds with the need of the general economy. There are indications that settlement has brought a much higher input of labour but a lower input of imported capital goods, i.e. a replacement of labour by imported types of capital. There are no precise data to confirm this statement quantitatively, but Table 20 provides some information which seems to substantiate this point. The farm expenses per 1,000 acres, taking into account agricultural supplies only, are roughly twice as high on large farms as in settlement areas, although the total production in the latter is certainly not smaller. The replacement of labour by capital leads therefore to rather high figures for the "value added".

A final factor of importance is the reduction of the foreign exchange component within farming expenses. On large farms, roughly 25—35 per cent of the total farm expenses is taken up by goods which have to be imported, while the foreign exchange percentage in settlement is but 14 to 18 per cent. In actual figures, the value of imported means of production (c. i. f. Mombasa) per 1,000 acres of settlement land is but half or a third of that in large farming areas. The difference is mainly due to lower inputs of machinery. Settlement areas consume little other imported goods than fertilizer, sprays, medicines, and some contractor ploughing and combining. If the future performance of settlement confirms that there is an increase in the marketed sales of exportable products, it can safely be assumed that the effect on the balance of payments will be positive.

c) Costs

The aforementioned benefits have to be compared with the costs in order to make an evaluation possible. According to the information given to the House of Representatives in July 1965, the total costs of the current programme, i.e. the One Million Acre Scheme, the IBRD/CDC Schemes and the land which is expected to be put to co-operative use, will eventually amount to £ 25.5 million, of which £ 18.5 million had been spent up to the end of the 1964/65 financial year. Most of the remainder will be spent in 1965/66 and the rest in small amounts up to 1968.

Of the total of £ 26.5 million, £ 21.6 million have been made available by the British Government, £ 1.6 million by the World Bank, £ 0.8 million by the Commonwealth Development Corporation, £1.2 million by the Federal Republic of Germany, and £ 1.4 million from local sources, i. e. the Agricultural Finance Corporation. The funds are loaned at a 6 per cent interest charge, except £ 9.7 million which have been given as a grant by the British Government to provide an element of subsidy in the land purchase — which is actually a compensation payment to the outgoing British farmer — and to cover some of the costs of planning, soil conservation and administration.

Altogether, total costs will reach £ 22 per acre and £ 805 per settler. However, only a part of that can be considered as development capital. £ 11.7 million of the total are for the land purchase. £ 5.5 million have been provided for administration and supervision. After these deductions have been made, about £ 7.7 million remain as development capital, i. e. £ 6.80 per acre and £ 230 per settler. This is a colossal injection of capital, considering the acreage concerned, far in excess, for instance, of what was spent on high potential African lands under the Swynnerton Plan.

The answer to the question as to whether or not resettlement is an economically justifiable venture — particularly in terms of the Treasury — depends on the repayment of the loans. Roughly £ 4 million were given to the settlers as land purchase loans and roughly £ 6 million distributed as development loans. In 1963/64, the first year of full settlement, the position with regard to the repayment of loans was fairly satisfactory — 77 per cent of the instalments due were paid. However, many of the repayments were not made from the income which the plots provided but were simply deducted from the down payments which the settlers had had to make when their applications were approved. In many other cases, income from employment elsewhere or from the farm in the former Reserve was used to pay the loan instalments. It is obvious that sooner or later the loan repayments will have to come from the productivity of the settlement holdings themselves and this will be possible only if the budget targets are reached.

Although settlement is likely to bring increased total production and sales, it is most unlikely to bring about the doubling of total production and sales within four or five years, which is nevertheless the basis for the servicing of the loans. Consequently, there is reason to be rather pessimistic about future loan repayments. The Department of Settlement is certainly taking more and more action against chronic non-payers. However, it must be realized that it will be extremely difficult to obtain payments from farms which have not been developed according to the budget. The Government of Kenya is definitely not in a position to evict a large percentage of the newly established settlers if, whether by their own fault or not, they have not been paying their commitments fully. It may well be revealed in the years to come that one third or even half of the loans distributed cannot be recovered and must be added to the costs of the resettlement programme.

In addition, other costs have to be considered:
One of the astonishing features of settlement is the apparent ease with which applicants can muster the required down-payments for Low Density or

Yeoman Schemes, which amount in all to no less than £ 700,000. Much of this money has undoubtedly been drawn out of the former Reserves. It may be assumed that the peasants of Kenya — within their traditional set-

Table 20. *Some estimates of the economic implications of production on large farms and on resettlement farms for the general economy*

£ per 1,000 acres

Large farms

Location Altitude Activities	Molo-Mau-Narok High Pyrethrum-Wheat-Sheep	Uasin Gishu Medium Wheat-Maize-Milk	Trans-Nzoia Medium Maize-Cattle
Total output	9,543	6,676	6,269
of which the export[1] content is:	3,876	2,186	1,236
Farm expenses[2]	5,018	4,057	3,429
of which the import[3] content is:	1,718	1,566	915
Value added[4]	4,525	2,619	2,841
of which the farmer's profit is:	65 p. c.	74 p. c.	72 p. c.

Settlement

Location Altitude Activities	Ainabkoi High Pyrethrum-Milk	Kaben Medium Milk-Maize	Mweiga High Milk-Potatoes	Ndalat Low Milk-Maize
Total output	5,767	6,190	10,483	5,501
of which the export[1] content is:	1,368	736	1,538	593
Farm expenses[5]	2,812	2,178	2,979	3,487
of which the import content is:	413	367	434	593
Value added[6]	2,958	3,512	7,504	2,029

[1] At farm level, including exports to other countries of the East African Common Market.

[2] Total farm expenses minus expenses for labour.

[3] At Mombasa.

[4] Total output minus farming expenses as defined before.

[5] Farm expenses plus wages paid to labour.

[6] Excluding wages paid to labour; including payments to contractors and loan instalments.

Source: Compiled out of information collected by MacArthur, J. D., Farm Economic Survey Unit, Nakuru, 1965.

ting — prefer to hoard the greater part of their savings in cattle and cash and that settlement has thus mobilized resources which would otherwise not have been used productively. Although this is certainly true to some extent, it must nevertheless be realized that the development of the former Reserves has suffered owing to the exodus of funds and especially of better farmers into settlement.

Another consequence of resettlement which must be kept in mind is the loss in revenue. Resettlement means a transfer of incomes from a few large farmers who pay some income tax to many smallholders who do not pay any income tax at all and who consume goods most of which are not taxable[1]. Some estimates to endorse this statement can be taken from Table 20. According to information provided by the Farm Economic Survey Unit, the profits of large farms amount to approximately 30 per cent of the total production value, which is almost equal to total sales. Assuming a tax rate of 6½ shillings in the £, the contribution of each £ 1,000 produced on large farms to the revenue of the Treasury amounts to £ 93, i.e. 9 per cent. Taking the already settled area as 780,000 acres and the previous production as £ 4/acre, we arrive at revenues of roughly £ 350,000. Only a fraction of these revenues will be forthcoming after settlement.

On the other hand, the Treasury will have to finance additional veterinary and agricultural services. It may be assumed — as a rough guide — that 250,000 acres of smallholdings require an annual expenditure of £ 50,000 for agricultural and veterinary services. The extension costs for the same acreage on large farms is likely to be less than a third of the above figure, i.e. for the 780,000 acres of settlement in the present scheme the Treasury will have to spend an additional annual sum of £ 100,000 for extension.

3. Evaluation

How does one strike a balance in assessing the difficulties, costs, and benefits of the current resettlement programme? There are some schemes where the budget targets will undoubtedly be reached even before the fourth year of settlement, where Settlement Officers are popular, where their advice is taken, where co-operatives work properly, where due loan instalments have been paid, and where future payments should offer no difficulty whatsoever because the production basis has been laid. There are also a few schemes on poor land, in a cold climate, run by poor Settlement Officers, occupied by disinterested settlers, where cattle was not regularly dipped and consequently died, where farmers have been slaughtering grade cows and selling the fencing and corrugated iron sheets provided to them on their development loans, where they do little work on their plots but look for employment elsewhere, where people do not understand what loans mean, where they will not believe what is explained to them, where they think they are cheated and sincerely feel that the land and the development funds should be given free of charge. And there are many more schemes which fit in somewhere in between these two extremes.

---

[1] In general, owners of large mixed farms do not pay much income tax, being in a position to write off most of the actual profit, but — more important — they consume goods which are highly taxed.

In trying to evaluate the resettlement programme as it stood in mid-1965, it must be said, firstly, that resettlement had to take place and that all the costs which might not be recovered are negligible compared with the economic losses which chaos would have brought to Kenya. Secondly, resettlement has worked amazingly well considering (1) the speed and (2) the scope of the undertaking, (3) the fact that most of the European and African personnel were hastily recruited and frequently lacked the professional or human qualifications for settlement, and (4) the administrative and political problems of a society in transition.

What L. H. Brown has written is undoubtedly true: "Leaving out land purchase, some £ 17 million could have been ... spent on extending and developing irrigation schemes, and on improved services and land consolidation in existing African areas, a comparable number of landless families could have been settled at least at equivalent standards and there would have been a greater upsurge in the overall production and standard of living." But this was not the alternative. Large parts of the former White Highlands had to be Africanised rapidly before and after Independence. The only alternatives to resettlement which might have been considered were (1) African large farms, (2) co-operative farms and (3) state farms. Considering the performance of African large farms in Kenya, of co-operative farms in East Africa and elsewhere, and considering the losses which are usually incurred by state farms, resettlement might have been, after all, the better choice in most locations, though certainly not in all [1].

However, one relevant set of questions remains: Would not an unplanned settlement have been more economic and politically even more advantageous? Was it worthwhile to provide the additional development loans and the manifold services supplied by the settlement staff? Did this mean throwing away good money after bad? Will not vigorous action against non-payers of loan-instalments create more bad feeling than expected? Would it not have been better to purchase the land, to compensate the outgoing British farmer with British funds, and to hand the land over to tribal committees for distribution without any specific loan schemes for development purposes?

Clearly this would have meant the expansion of the type of farming prevalent in the poorer parts of the former Reserves. There would have

[1] One location where one cannot but have the greatest misgivings, is the Kinangop, a high plateau of 200,000 acres west of the Aberdares, with mainly sour, heavy soils in a frostprone climate, where Kikuyu have been settled. Referring to the Kinangop L. H. Brown writes: "This type of settlement ... may in the end rebound on its advocates by producing a body of dissatisfied paupers unable to pay their settlement dues and who have to be evicted. This is an area overdue for a critical reappraisal". In the eyes of the British settlement personnel the schemes on the Kinangop were an unavoidable political sacrifice.

been no grade cattle, little milk production and but a few acres of cash crops in the area of unplanned settlement. The marketed production would have dropped a great deal to perhaps a third or half of its former value. However, roughly £ 17 million would have been saved and could have been invested productively elsewhere.

The correct answer to these questions will certainly depend on the future performance of the resettlement schemes. If production increases significantly and if most of the loans are repaid in time, then the present form of resettlement might well be considered to have been the better choice. It is true that the present information shows some increase in production and some improvements in loan repayments on a limited number of schemes. But it is not yet possible to assess the position for all schemes. In any case, the economic impact of resettlement should not be overrated. We fully agree with L. H. BROWN when he says that "the steam is still in the kettle, because the basic source of heat, the 3 per cent per annum increase in overall population, is still beneath it. Since 1962 one may calculate that the better part of another million people have been added to Kenya's population, or over 100,000 families, serveral times as many as the number of new families settled on all the schemes . . . After spending £ 25 million ostensibly to relieve population pressures and land hunger, the overall position in Kenya in 1965 is worse than it was in 1962".

VI. African Large Farms in the Former Scheduled Areas

The other important move towards the Africanising of the former "White Highlands" has been the support given to individual Africans in the purchase of large farms within the still intact belt of large farming which stretches from Nairobi to Mt. Elgon [1]. The large farming areas of Kenya were, up to a few years ago, controlled by legislation, which prevented the transfer of land to persons other than Europeans. In the African areas this legislation had its counterpart — although of little practical re-

[1] Another step towards the Africanisation of the Highlands, which is rarely recognized as such, is the increase in squatter cultivation. Farm workers have always cultivated plots in order to grow their subsistence food. In recent times, squatter cultivation has increased rapidly. Relations and friends establish themselves at the farm worker's place and take the odd corner of the large farm into cultivation. On some farms the acreage taken by the squatters amounts to one quarter of the total. The farm owner, assuming that he will be bought out sooner or later, thinks first of his personal security and therefore shies away from decisive action against squatters. In some places the increase in squatter cultivation actually amounts to an unofficial kind of unplanned settlement. The subsistence value of squatter cultivation is certainly important.

levance — which prevented land from passing into the hands of persons other than Africans. The purpose of this measure was to create land stability in order to make farming in the Highlands attractive to European settlers, a thing which — in former times — was considered to be essential for the economic development of the Colony. The Royal Commission on East Africa, as early as 1953—55, advocated the removal of these racial restrictions on transfers of land. However, its recommendations were not accepted until six years later at the Lancaster House Conference in 1961. Thereafter everything had to take place in a hurry. Since 1962 increasing numbers of Africans have been buying large farms and have been helped to do so. (See Table 21).

Table 21. *Analysis of loans approved by the Land Bank in 1963 and 1964 to assist in the purchase of land and permanent improvements*

	Value of Loans approved in £		No. of Loans approved		Acreage involved	
	1963 £	1964 £	1963 No.	1964 No.	1963 acres	1964 acres
Total	1,352,625	1,224,130	238	277	469,854	306,470
Per cent given to	%	%	%	%	%	%
Africans	39	73	21	80	11	50
Asians	5	2	6	2	5	1
Europeans	56	25	73	18	84	48

Source: Land Bank, Nairobi.

1. Aims, Institutions and Methods

a) Aims

The main objective behind public support for African buyers of large farms are generally the same as those of settlement:

— there should be tangible evidence that large parts of the former White Highlands are now in African hands, and this should lead to an easing of the tensions which have built up over the land question.

Other objectives have also to be considered:

— The Africanisation of the large farm economy should lead to a situation where the economic power, which goes along with larger properties, is wielded by members of the majority race whose loyality lies with Kenya. Their profits are expected to stay in the country, while those of the minority groups are likely to be transferred elsewhere.

— After the Lancaster House Conference, enough buyers of land, other than Africans or Asians, could not be expected to come forward. The buying of land by Indians would have replaced one racial problem by another, and could therefore hardly be promoted by public funds [1]. If the Land Bank had not pursued a policy of support for African buyers, land values would have dropped even more and there might have been a pronounced running down of the whole large farm economy.

In order to establish Africans on large farms, public funds were made available as loans provided by the Land Bank [2], the Agricultural Finance Corporation [3] and — in some cases only — the Ministry of Settlement. Private banks are involved in lending working capital to some new farmers of repute.

The total area occupied by Europeans and a few Asians in Kenya in 1960 amounted to 7.6 million acres. Of this, 7.25 million acres were situated in the former White Highlands. 3.5 million acres are taken by ranches and approximately 450,000 acres by coffee, sisal, tea and wattle. African ownership in these capital intensive activities has been limited to a few cases. The interest of the African buyers focuses on the mixed farms, which took up roughly 3.4 million acres in 1960. Of this, Settlement has taken over approximately 1.2 million acres, leaving a mixed farming area of approximately 2.2 million acres at the end of 1964. There were, in March 1965, 524 new African large scale farmers in the area, occupying approximately 500,000 acres, that is 22 per cent of the remaining mixed farming area. The racial change in landownership, which has become possible owing to the dropping of legal barriers, has been helped by a number of factors in addition to the loans provided by public institutions:

- The political change led to a drop in land prices. In 1964 prices of farms were about 75 per cent of what was paid before the Lancaster House Conference.
- Land prices in the former Reserves, particularly in Central Province, have always been higher than in the Highlands. Land of comparable quality which changed hands for £ 30—35 per acre in the Highlands received prices of £ 100—120 in the Reserves. The drop in land values in the Highlands after the Lancaster House Conference made the difference in prices even more pronounced.

---

[1] It is assumed that up to the end of 1964, 60—70 European farms in Rift Valley Province have been bought by Asians. Asian interest — so the rumour goes — slowed down when some important African politicians warned that the Emergency had not been fought merely in order that there will be the "Brown Highlands" (with Asians) instead of the former "White Highlands" (with Europeans).

[2] The Land Bank operates with a certain amount of Kenya Government funds, a lot of British aid money and some local commercial money. It lends money to individuals or co-operative groups or companies for buying land against good security.

[3] The Agricultural Finance Corporation took over the function of providing loans from the Central Agricultural Board, which used to cover the former Scheduled areas.

— There were a substantial number of run down farms in the Highlands which changed hands for very little money indeed.
— Cash crop development in the former Reserves had created surpluses which were used for buying land. An amazing number of family groups were in a position to mobilize substantial hoardings of cash and cattle in order to exploit the unique chance of buying land cheaply.
— In a few cases, coffee-co-operatives mobilized their resources and bought European coffee estates.

Table 22. *Analysis of size, value and returns of large scale African farming in the Rift Valley Province* [1]

	Kitale District	Eldoret District	Kericho District	Nakuru District	Total
Numbers of farms bought by Africans	88	220	24	132	464
Total acreage	57,200	320,000	9,300	196,000	482,500
Number of farms over 2000 acres	4	10	—	11	25
1000—2000 acres	19	127	1	11	138
500—1000 acres	21	30	3	29	83
under 500 acres	44	53	20	81	198
Value of land and other assets on farms bought by Africans, in £'000	572	2,200	55	2,600	5,427
Estimate of present annual gross return in £'000	75	550	24	450	1,100
Present gross return in per cent of the gross return which could be expected under average management conditions[2]	30	30	50	30	35

[1] Most of the former "White Highlands" are situated in Rift Valley Province.

[2] This estimate is not based on a systematic survey. It represents the personal impression of European farmers and Agricultural Officers who have known the area for some time and who compare the present situation with the one on average European farms.

Source: Department of Agriculture, Rift Valley Province, Annual Report — 1964, Nakuru, Mimeo.

The reasons for buying land are rather different from those experienced in former times, when individual Europeans with enough capital backing endeavoured to farm in order to make a profit. As a rule, the new African farmers are motivated only to a minor degree by the profit motive. More important is the idea — which is probably correct — that land values are low and that to buy land now should yield a handsome dividend. Paramount, however, seems to be the traditional attitude that to own land is a value in itself, that it gives social standing to the owner, as well as social prestige and security to future generations of his family.

b) Institutions

There are a number of different ways in which the racial change in landownership is taking place:

So-called "*Assisted Owners*" have been established on larger subdivisions of ex-European farms. Up to the end of 1964 there were 118 of them settled on about 60 former European farms. The scheme was originally supposed to qualify for funds from the World Bank (IBRD) and the Colonial Development Corporation (CDC), but it never reached the specifications required by these agencies and had to be incorporated into Settlement. Thus the Assisted Owners are settled with funds given for settlement purposes, but outside the specific settlement areas.

Within the purchase programme of European farms for settlement financed by the British Government, provisions were made for the purchase of some farms outside the settlement areas, which were owned by so-called "*Compassionate Cases*", i.e. farmers of ill health, old age etc. The purchase was handled by the Central Land Board. The farms were sold to African individuals mainly, in some cases to partnerships, companies and co-operatives, on the basis of only 10 per cent of the price as the initial deposit, the remaining 90 per cent being repayable in equal instalments of principal and interest over 20 years. Most of the farms were in an extremely poor condition.

In a substantial number of cases, *private arrangements* have led to some degree of Africanisation: one or more Africans have entered into some kind of partnership with the European farmer. The basic principle behind this is that the farm is purchased step by step by the new partners, the price being paid partly out of the forthcoming profits. Knowledge about these arrangements is rather scarce.

Most important, perhaps, not as regards numbers but as regards the acreage and the production potential, are *purchases arranged with the Land Bank*. Those who buy their farms with a Land Bank loan have to take the following steps: (1) obtaining a Land Bank loan, (2) obtaining a loan from the Agricultural Finance Corporation, (3) obtaining the consent of the Divisional Land Board. Tho Board can withhold approval on the grounds that the price is too high or the applicant lacks sufficient farming experience. But the whole transaction hinges on the Land Bank loan. When this has been granted it is most unusual for any of the rest to be withheld.

c) Methods

The methods of take-over vary according to the institutions concerned, the number of buyers, and the amount of capital they possess. The Bank has, for some time, tried to insist that a purchaser has, in cash or easily convertible assets, 50 per cent of the purchase price of the land and permanent improvements. As evidence of this the applicant is required to furnish a certificate from his bankers or lawyers, stating the amount of cash held. In cases of native livestock held in the former Reserve, the Bank also endeavours to obtain a certificate from the local chief or headman.

Table 24 shows the ideal case of the purchase of a 1000-acre farm supported by the Land Bank. Land and permanent improvements are purchased at £ 9 per acre, i.e. £ 9,000 for the whole farm. In addition, £ 5,400 have to be paid for the movable assets (cattle, machinery etc.). This, together with several charges, adds up to £ 14,692. Of this, £ 7,200 is normally carried by a Land Bank loan and £ 4,300 by a loan from the Agricultural Finance Corporation. These two loans add up to £ 11,500, leaving a balance of £ 3,278 (including loan charges) which, theoretically, has to be carried by the applicant. Also, the buyer is expected to have approximately £ 1,200 for running expenses. Thus the capital requirements add up to approximately £ 4,500. With the assistance of crop finance for planting expenses, which is generally provided for large farms in Kenya, he should then have a reasonable start. The Land Bank loan has to be repaid over 20 years and the loan from the Agricultural Finance Corporation over 10 years. In both cases 6.5 per cent interests have to be met.

The procedure as described above is, however, the ideal and not the norm. The shortage of capital has led to rather complicated arrangements in farm management and ownership. LINDSELL distinguishes the following methods of acquisition:

Acquisition by one person using only his own assets. This is a very rare case. These buyers are usually people of some knowledge who can solve their practical difficulties without undue trouble.

Acquisition by more than one person using assets solely their own. There are several, though not many, cases like this. They should, theoretically, present no difficulties if the applicants came as genuine partners. However, not only are the intricacies of partnership law quite unknown to them, but it also happens that they have no real intention of acting as partners. Several cases have occurred where two or three people have joined together to obtain a farm, but have immediately started to farm in separate units.

Acquisition by one or more persons using assets provided by many others — family groups or friends, numbering 25 and more. This is generally the case. For practical purposes the loan authorities accept this proposition provided that the lenders acknowledge in writing that the loans are not repayable for a considerable number of years, so as to avoid a heavy debt resting on the farmer from the outset. Of course it is stressed to the lenders that the loan gives them no right whatsoever to the farm. It would, however, be most unrealistic to imagine that this really sinks in. The lenders almost certainly have the conviction that the new farm-owners are merely trustees for the group and that, if the need or the wish arises, they can all live on the farm. There is virtual certainty of future legal trouble in the event of the death or the absence of one of the partners.

Acquisition by a co-operative. There are some cases where a great number of workers who were employed by the former European owner have formed a co-operative and bought the farm with a down-payment of about 10 per cent. In other cases, an outside group co-operatively buys the farm. The number of co-operatives — which offer many legal advantages compared with other methods of acquisition — is kept low because the Co-operative Department has ruled that a new society may acquire a farm only when every member intends to live and work on the farm. In fact, in most groups this is not the immediate intention of all members, who often represent a larger family group.

Acquisition by a limited liability company formed by the buyers. There are a number of such cases, but this is an unsuitable form of organisation for the groups concerned. The relevant law was devised for an industrialized society. Where companies have been formed, it is likely that serious difficulties will arise over compliance with statuary requirements, accounts etc.

2. Difficulties, Benefits and Costs

a) Difficulties

Within the few years since Africans began to take over large mixed farms it has become evident that a large proportion of them are not doing well. There is, as always with social phenomena, a very wide range in the performances. Some of the African large-scale farmers are running their enterprise as well as would any European settler. An even greater number of farms shows the promise of being run successfully. There are notable exceptions among the newly established co-operatives and farming companies. Those who bought their land early on, in 1962 and 1963, were usually better supplied with capital and had more business experience than those who joined more recently. The Kikuyu farmers, having had longer experience than others in commercial life, are as a rule more easily adaptable to managing large farms than are members of the Kalenjin tribes. However, efficient as the exceptional cases might be, the performance of a great number of farmers is poor. According to the estimates of Table 23, which may be on the pessimistic side, only a small fraction (5 to 10 per cent) of the farms is properly run. Particularly bad is the situation of those who took over farms as "compassionate cases" or by private arrangement. The position of those who bought their land with a Land Bank loan is, on the average, somewhat better. Most of the difficulties can be traced back to a few important, interacting shortcomings:

α) Lack of Capital

There is a desperate shortage of ready cash for proper economic operations. The Land Bank demands that the applicants possess 50 per cent of the price of the land and the fixed assets in ready cash or readily available assets. The buyer then usually gets a loan of 70—80 per cent of the land price, but sometimes only 60 per cent; private resources are expected to provide the balance. The farmer is, however, immediately short of cash for the purchase, let alone for running expenses. The fullest possible use is made of all other forms of credit available, which in due course leads to a heavy burden of debt. The actual situation is much more acute than might be assumed by taking the above figures at face value. We have to take into account the following facts:

- The cash and the assets produced usually do not belong solely to the applicant but have partly been borrowed from friends and relations for the purpose of satisfying the Bank's requirements. As soon as the loan has been approved, the "borrowed" element is returned to the lender.
- Private Banks are reluctant to provide the loans they used to give to European farmers. They consider most of the new owners to be too much of a financial risk.
- Some of the available funds are not used to develop the farm, but to buy more land because it can be bought cheaply.
- Many of the new farmers are reluctant to dispose of all extraneous assets. They tend to keep their farm in the former Reserve. This, of course, increases the shortage of capital and leads to divided interests and absenteeism.
- The lack of capital leads to insufficient stocking with grade cattle. The new farmers tend, therefore, to stock their grazing with low grade cattle from the former Reserve, mostly belonging to relations and shadow partners. The introduction of low grade cattle, together with insufficient dipping, in turn endangers the mixed farming system, which is built on the efficient husbandry of grade stock.
- The desperate lack of capital is made even worse by the tight capital situation generally. Ten years ago, when British settlers were becoming established, Land Bank loans were available at 4.5 and 5.5 per cent interest. Now the interest is 6.5 per cent. Moreover, the African large farmers of 1965 have to repay their loans in 20 years; formerly it was 30 years. It must not be overlooked that the new African large farmer is facing a more difficult situation than did his British predecessor 10 years ago.

β) Lack of Appropriate Institutions

The fact that a great number of people have to group themselves together in order to produce the minimum capital requirement of the lending agencies causes serious difficulties in farm management. Mismanagement arises from the break-up of partnerships which have been newly formed. This makes it difficult to advance loans. Family groups and friends, standing behind the buyer as shadow partners, invade the farm as of right. A multiplicity of partners can be found on some farms, each expecting grazing and residential rights and firmly objecting to proper planning and expenditure on development and maintenance. Illegal subdivisions are carried out. Farms which are entities from the standpoint of legal ownership continue to be cultivated in practice as if they were a number of different holdings. In some cases up to forty different herds, each belonging to one "partner", have been found grazing as distinct units. The land is sprinkled with huts. Some farms have been sold to co-operatives which not only lack capital and know-how but also leadership and discipline. Some of them have only been established in order to obtain preferential treatment with loans. It is difficult for the Agricultural Officer or the Land Bank agent to find out with whom he should be dealing and who has the authority to make decisions. Very few farms are actually run by one man or manager. There is evidence of a kind of infiltration by members of partnerships, companies and co-operatives who establish some sort of squatter cultivation and whose low grade cattle introduce ticks and disease. All in all, the danger exists that some of the land bought by Africans is reverting to the mediocre subsistence agriculture prevalent in the poorer parts of the former Reserves, with the hoe as the predominant tool.

Table 23. *Analysis of take-over and management standards of African large farms in Rift Valley Province*

Number of large farms bought by Africans	Arrangement of take-over	Source of[1] finance	Management standards[2] per cent of farms in classes: good	fair	bad	Ownership—number of Individuals[3] & Partners	Co-operatives	Companies
118	Assisted owners	30 p. c. buyer 90 p.c. & more HMG[4]	5	75	20	118	—	—
170	Sales arranged with the Land Bank and the Agricultural Finance Corporation	50 p.c. buyer 50—80 p.c. L. B.[5] 50 p.c. AFC[6]	10	75	15	93	77	
133	Compassionate purchases	30 p.c. buyer 90 p.c. LB[5] 50 p.c. AFC[6]	5	35	60	71	4	58
100	Private arrangements, mainly liaisons	Rental through lawyer until purchase funds are available		40	60	?	?	?

[1] The percentage figures refer to the purchase price for land and fixed assets on the farm (buildings, water supply etc.). They do not include movable assets (cattle, machinery etc.).

[2] Management standards are assessed using the level of farming of new farmers. If they were assessed using the standard of European farming, the percentage in the top category would be extremely small and that in the bottom category much greater.

[3] Almost invariably the individual owner has a group of shadow backers behind him, who have no legal status on the farm concerned.

[4] HMG — Her Majesty's Government (British Government Funds).

[5] L. B. — Land Bank.

[6] AFC — Agricultural Finance Corporation.

Source: Unpublished Estimates of Personnel of the Kenya Government.

γ) Lack of Know-how and Drive

More important than the above shortcomings is the inadequate background of the new large farmer. His previous economic experience has been mainly of smallholder farming, with a strong subsistence element. He has little understanding of the harsh economic pressure of a formal loan structure with its interest element and strict repayment covenants. The new farmers are handicapped by a lack of formal education. LINDSELL estimates that less than 25 per cent speak English, possibly 50 per cent are able to write with ease, and a few cannot write or read at all. It can hardly be expected that many of them will be able to produce income and expenditure accounts. Even those who have had successful semi-commercial farms in the former Reserves are not accustomed to the details and peculiarities of large-scale arable farming. The full importance of the timing of mechanised ploughing and planting operations is not clear to them. The need for proper animal husbandry is not sufficiently realised; finance for filling the dips with adequate concentrations of chemicals is lacking, and consequently disease is spreading. 1,300 head of grade cattle are said to have died from tickborne disease in 1964 on African large farms in the Kitale and Eldoret Districts alone. Furthermore, there is some evidence of scrub bulls, overgrazing and destruction of fencing.

A certain number of the new large farmers feel insecure and therefore neither ask for advice nor pay attention when it is offered. "We are opposed to those people who come only to find fault and to discourage our endeavours" is one of their complaints, handed to the President in January 1965. An amazing number of those who invested all they had and could borrow show little ambition to improve and to work as hard as is necessary to get out of a desperate economic situation. Not only do several of them lack the know-how and capital, but also the necessary drive. However, public demand is clamorous for more and cheaper Government loans. African large farmers claim that the Land Bank evaluation officers are particularly "friendly" to Europeans who sell and set the value of the farms too high, that the Land Bank requires too high a down-payment, and that it is unfair that those who were not able to pay back their dues have had their names circulated to other financing organisations. The trouble is that the new African farmers have placed themselves in the position of capitalistic entrepreneurs without realising the stringency of economic duties which most of the capitalists in Europe, North America and Japan impose upon themselves in order to morally justify their enhanced position.

b) Benefits and Costs

a) Benefits

There is, fortunately, still a belt of large farming in the Highlands of Kenya, but there is no longer a White Belt and this is undoubtedly an important improvement in Kenya's present political situation. Some 500,000 acres have been Africanised and some 500 Africans possess land which used to be owned by Europeans. It is hoped that future profits of the

Africanised large farms will stay in the country, and there are some indications that African large farmers will not imitate European standards of living, but will stick to the modest ways of living prevalent among the better farmers in the former Reserves. Furthermore, it must be pointed out that support for African large farms is the cheapest way of achieving the Africanisation of the land; that, as with other forms of settlement, it is mobilising private rural capital and developing managerial talent.

β) Costs

However, the costs and losses to the economy in general and to the Treasury in particular are already substantial. We can list the following:

Table 24. *Planning figures of the Land Bank for the Africanisation of large mixed farms* [1]

		£ per 1000-acre farm
Price of land and permanent improvements		9,000
Costs of loose assets and development expenditure		5,400
Various charges for the transfer		292
Total		14,692
Less maximum Land Bank loan	7,200	
Agricultural Fin. Corp. loan	4,300	
	11,500	
Various charges on loans		106
Purchase capital required from the buyer		3,298
Working capital required from the buyer		1,200
Capital requirement for a buyer of a large farm		4,498

[1] These figures represent a model case for the purchase of a large farm up to 1965.

Source: Land Bank, Nairobi.

— *A drop in production.* It is estimated by the Agricultural Department that the large farms taken over by Africans produce only one third of the gross return which could be expected under normal conditions and proper management. (See Table 23). It has, of course, to be taken into account that many, if not most, of the farms which changed hands had been allowed to run down and were far below the average standards of the European settler community. This is particularly true of the "Compassionate Cases".

— *Losses of loan funds and arrears in repayments.* In a certain number of cases the farms are already on the verge of bankruptcy and it will not be easy to set them on their feet again. Here and there instalments are met by the sale of assets or from non-farm income. This, however, offers no lasting solution. Although the Land Bank money, backed by the security of land mortgages, may be considered safe, on many farms a considerable share of the Agricultural Finance Corporation investment, generally secured by mortgages on chattels, is in jeopardy. In addition, there are the effects of the prolonged drought in 1965 which put many of the new farmers in a difficult position. The Government must take into account that some of the funds loaned out are already lost and that in a great number of cases the Bank might be asked for a moratorium of several years on all interests and capital repayments.

— *Loss of revenues.* In addition, the Treasury will have to reckon with a loss of tax revenues. The income tax from the farms concerned will be almost nil. In addition, there is the loss of revenue owing to the reduction in marketed production and thus in secondary income.
— *Repercussions on the large farming community as a whole.* It is sometimes assumed that the African large farmer will provide a "social umbrella" for the remaining European farmers and that the mingling of African and European properties makes the presence of an important European farming community more acceptable in political terms. However, the development seems to go to the other way. There is frustration owing to mounting stock thefts, trespassing and stealing — of maize in particular — which can partly be explained by the rapid growth of squatter societies. Furthermore, a section of the African large farmers is developing some degree of animosity towards European farmers whose technical and economic performance contrasts with their own and whose commercial and credit affairs seem to run more smoothly.
— *Repercussions on the non-farming community.* Much, if not most, of the non-farming economy of Kenya is built on secondary activities relying on the primary income generated in agriculture. This ist particularly true of the large farming sector which functions with a high degree of division of labour. Any lowering in the efficiency of large farm production is therefore likely to have significant repercussions on urban activities and employment.
— *Higher costs for agricultural services.* Most unfortunately the Government of Kenya is not in a position to serve management orders to all those large farmers, — Europeans, Asians or Africans — who fail to repay their loans. Were this to happen, there would be strong feeling among those Africans who became dispossessed; there would be strong political pressure. The Government will therefore have no choice but to provide additional agricultural personnel to advise and supervise the new farmers. This is likely to be very expensive indeed. Up to now, the costs of the Agricultural and Veterinary personnel in the Scheduled areas have been rather low, but with increasing Africanisation this will change. By 1965 the Treasury had already had to allot a sum of £ 30,000 per annum for additional extension services to African large farms.

3. Evaluation

It is most important for Kenya, and in particular for the large farming community, that Independence coincided with several major achievements in applied agricultural research. (Rust-resistant varieties of wheat, the introduction of hybrid maize, the development of protein-rich clover mixtures for leys, etc.) Furthermore, large investments were made between 1955 and 1961 which are now in their productive years. Both these factors stimulated the large farm economy to such an extent that the loss of land to settlement and the drop of production on many African large farms is hardly discernible in the 1963 and 1964 output figures of the large farming sector as a whole. The loss of the mixed farm economy was more than offset by increases in the plantation industry. The fact that there has not yet occurred a statistically recognizable major break-down in the large farm economy should not, however, obscure the real situation. The problems are coming up and they are likely to be very expensive ones.

The benefits of public support for African large farms go to a small group of people who are in the fortunate position of having access to large loans. A certain percentage of them are unlikely to make efficient use of their privileged position. However, they can be quite certain that even a poor economic performance would hardly endanger their newly gained wealth. Their political position is already a rather powerful one. There would be a general racially-based outcry of opposition if the Land Bank or the Agricultural Finance Corporation were to try and to oust more and more defaulters. Up to early 1965 not more than two African large farms had been taken over and placed under Government control, and this is certainly not enough to rectify the situation.

VII. An Economic Comparison

The discussion of the various aspects of African agricultural development policies, (1) administration and extension, (2) tea-development, (3) grazing development, (4) settlement on dry and irrigated land, (5) resettlement of European land and (6) African large farms leads to the question: how do they compare with each other? In this respect two criteria seem to be of decisive importance, firstly the increase in marketed production and secondly the increase in revenues to the Treasury. The financial results in terms of the Treasury are of particular importance for Kenya. Internal revenues are scarce and most of the additional development effort will have to be paid for out of loans. In the long run, services and schemes for African farmers must therefore create the very revenue with which the Treasury or related bodies can serve the loans and finance additional efforts.

In Table 25 a tentative effort has been made to compare the economic returns of investment in different development measures. The estimate does not include investments in grazing schemes and financial support for African large farms, because efforts in these directions have resulted in additional marketed production and additional revenues in exceptional cases only. The comparison has therefore been restricted to more successful or promising efforts, i.e. to (A) Smallholder tea development, (B) Agricultural administration, extension and related services in three districts, (C) Irrigation development, excluding Galole, the Tana River pilot scheme, (D) Low Density Resettlement and (E) new settlement on rain-fed land.

As a yardstick the "internal rate of return" has been used. This method is usually applied by the World Bank in order to evaluate capital development projects [1]. The "internal rate of return" shows the rate of interest which an investment yields. The procedure is fairly simple: the excess of receipts over payments of every year — i.e. the "annual operating savings" — are collected or estimated. Thereafter the actual rate of interest at which the discounted receipts are equal or close to the initial investment is calculated. In Table 25 the additional marketed production and the revenues of the Treasury have been treated as receipts. The revenues of the Treasury have been estimated as follows: operational savings of the scheme, plus 12 per cent of the value of the additional marketed produc-

Table 25. *A comparison of the economic returns of several development measures in Kenya agriculture*

Nature of investment	The internal rate of return [1] in terms of (I) additional marketed production	(II) revenues to the Treasury [2]
A. Smallholder Tea Development [3]	50	23
B. Agricultural Administration and Extension		
Nyeri District	40	10
Elgeyo-Marakwet District	35	6
Central Nyanza District	12 [4]	nil
All three districts	29	5
C. Irrigation Development		
Mwea-Tebere Scheme	22	8
Mwea-Tebere and Perkerra Scheme together	18	6
D. Low Density Resettlement		
90 per cent fulfilment of targets	9	4
70 per cent fulfilment of targets	6	2
E. New Settlement on Rain-fed Land	4	1

[1] There are numerous pitfalls in the use of the internal rate of return. (See: HEISTER, M.: Rentabilitätsanalyse von Investitionen, Köln-Opladen 1962). In this table the internal rate of return is giving a rough indication of the relative advantage of one form on investment as against the other.

[2] And related bodies. Contributions to the County Councils have not been included in order to improve the comparability of the estimate.

[3] Field operations only.

[4] Due to assumptions which had to be made. See Appendix V, B.

[5] A great number of assumptions had been made in order to make the estimates somewhat comparable. See Appendix V.

[1] See: SCHAEFER-KEHNERT, W.: Kriterien der wirtschaftlichen Beurteilung landwirtschaftlicher Entwicklungsprojekte. Agrarwirtschaft, No. 1, Jan. 1965. — TAK, H. G. v. d.: The Evaluation of Agricultural Projects: A Study of some Economic and Financial Aspects. IBRD Rep. EC-128, Washington 1964. — Harvard Business School: Planning Capital Acquisitions, EA-C 402, 1956. — SCHNEIDER, E.: Wirtschaftlichkeitsrechnung, Tübingen-Zürich 1964, p. 10. — HEISTER, M.: Rentabilitätsanalysen von Investitionen, Köln-Opladen 1962, p. 90.

tion[1]. As regards the other assumptions and data underlying the estimates made in Table 25, see Appendix V.

It is obvious that the estimates of Table 25 have to be interpreted with care. The assumption that the Treasury receives 12 per cent of the marketed production by taxation on produce bought by the farmers is rather arbitrary. However, such an assumption had to be made in order to compare investments in services with those in schemes. The costs and returns of agricultural administration and extension exclude the three districts, the overheads of which are substantial. They are based on fragmentary information. The costs and returns of smallholder tea development do not reflect the actual performance in the field as is the case with the other measures. They are based on the planning data of the KTDA for the 25,500 acre scheme. The returns of the Low Density Resettlement, again, are based on calculations for future resettlement. They only partly reflect actual information about the current schemes.

The shortcomings of the data behind Table 25 are rather large. A great number of assumptions had to be made in order to reach some degree of comparability. We consider nevertheless that the result is a useful contribution to the assessment of the relative advantage of one development measure compared with others:

— Smallholder tea development seems to be much more economic than any other proposition. The investment of the KTDA provides a high return in additional marketed production to the benefit of both the KTDA and the Treasury. The great economic advantage of the tea scheme lies in the fact that the public outlay is recuperated by a special cess on the green leaf. However, experience shows that hardly any planning target is ever reached, so the final outcome of the tea scheme might perhaps not be as good as was assumed in 1965. Yet this scheme would have to run at substantially higher costs and with lower yields before it would lose its position as the most economic venture in Kenya's present agricultural set-up.
— The returns of investment in agricultural administration and extension services, (including land consolidation, small loans etc.), when measured in terms of additional marketed production in Nyeri District (mainly coffee) and Elgeyo-Marakwet (mainly pyrethrum, milk, wool), are rather close to those of tea development. There is certainly no clear cause-effect relationship between the investments in services and the additional marketed production. Yet the coincidence of a large increase in sales with the increase in governmental agricultural activities makes it likely that the two phenomena are closely related and that public investments were rather effective in inducing additional agricultural production, in two of the three selected districts. However, the re-

[1] World Bank estimate for revenues from additional primary income in Tanganyika before independence. See: IBRD: The Economic Development of Tanganyika, Dar es Salaam 1960.

turns in terms of the Treasury are not as good. There is no direct recuperation of the public outlay except in the small tax on exported coffee. The treasury must rely on the revenues coming from the purchase of taxed goods and on secondary income generated by the farmers' additional selling and buying. The estimate in Table 25 implies that investments in agricultural services are likely to lead to well-to-do smallholders and an empty Treasury. In this connection it should not be overlooked, that the estimates are based on local (district) costs only. The outcome would be less favorable for services of this nature, if we include the heavy expenditure for overheads.

— The returns of irrigation development on the other hand have been remarkably high, compared with similar efforts elsewhere, largely owing to the fact that the Mwea-Tebere Scheme has brought a substantial increase in marketed production as well as considerable operational savings. The returns are high, even if we lump the poor results of the Perkerra Scheme together with the good ones of Mwea.

— Resettlement in Low Density Schemes, even if the planning targets are reached up to 90 per cent, shows a lower return, i.e. the investment is high in relation to the additional marketed production that can be expected and to the loan repayments. If only 70 per cent of the planning targets is fulfilled with regard to marketed production and loan repayment, the situation becomes even less favourable. The outcome of High Density Schemes is likely to be even less favourable. Clearly the estimates indicate that resettlement in the Highlands cannot be justified by its direct economic benefits to the nation. Resettlement must primarily be justified in terms of its social and political benefits, which, of course, by contributing to political stability, have important economic implications.

— New settlements on rain-fed land show a lower return than resettlement as to the additional marketed produce, although land which hitherto has produced nothing is brought to production. The financial result to the Treasury is likely to be even worse.

Whether the results of the past decade, summarized in Table 25, can be used to assess future results of similar approaches is of course open to doubt. Political independence and the rapid social evolution going along with it, changed the conditions for agricultural development. Approaches which failed in the past may succeed in the future and vice versa. However two points seem to be of general validity within the present African scene: (1) settlement schemes are very expensive indeed and are likely to yield acceptable rates of return under very favorable conditions only; (2) the mobilisation of great numbers of smallholders with the help of a close net of rural development services is likely to yield acceptable revenues to the Treasury only if there are substantial cesses on the additional produce or substantial taxes on the land.

D. Changes Since Independence

I. Africanisation

The drastic changes in Kenya's agricultural policy, i.e. the land tenure reform, the cash crop policy for smallholders, the resettlement schemes, and the support for African large farms, were already under way before political independence was obtained. Independence brought no radical reorientation — as it did for instance in Tanzania — but the intensification of already existing efforts. The institutional framework of the agricultural administration remained largely the same, the only significant change being the amalgamation of the different bodies dealing with the former Scheduled areas and the former Reserves.

Independence has, however, sped up the process of Africanisation in many ways: the increasing share of marketed produce sold by smallholders, the settlement of 26,000 African families on former European land, and the growing number of African large farmers. There is an active loan policy to promote African enterprise. Loans will not be issued to non-citizens and agricultural land will not be sold to them, except in special circumstances. Particularly conspicuous is the Africanisation in Government and related bodies. The replacement of European by African officers has taken perhaps somewhat less time than was expected, but it has definitely not been as abrupt as it was in Tanzania in 1962—63.

The Africanisation of Kenya's agricultural services started earlier and on a broader basis than in neighbouring Tanzania, and this made the change after Independence far less pronounced. The staffing situation is difficult partly because of the resignations of British officers under the compensation scheme and partly because of the need to supply technical staff to the Department of Settlement and the Kenya Tea Development Authority. However, the training of African officers has produced most of the replacements. The newly promoted officers will need some time to adjust themselves to their positions and a certain percentage — those who are overpaid in relation to their performance — will have to be weeded out. But generally we can say that, in spite of the difficulties of a transitional period, the Agricultural Department and the other bodies concerned with agriculture continue to remain intact and to perform their routine work as usual.

Nevertheless, a number of difficulties which are typical of any transitional period have not yet been solved:

— The situation is marked by the frequent movement of personnel. A field officer in the 1950's usually remained in one location for $2^1/_2$ years. In 1964—65 the average length of stay in one location was down to an estimated six months. This is obviously a serious handicap for any sustained work.

— The approach of Independence was accompanied in Kenya, as elsewhere in East Africa, by some degree of erosion of the Government's authority. Water rates, grazing fees and loan instalments were no longer paid as they should have been. In some locations loans had been considered as a bonus for "Uhuru" (Independence) and repayments became neglected. In 1965 the Government was trying hard to remedy the situation by bringing defaulters to court. The solution to the problem is certainly not an easy one.
— Some local politicians have not yet realized that the Civil Service, whose members are not expected to interfere in politics [1], is no longer the agent of the British Colonial power but the servant of Kenya's African Government. In some locations the division of responsibilities between administrators and politicians has not yet been sorted out properly.

Altogether, Africanisation in Kenya is going ahead with far less emotionalism than in Tanzania a few years ago. In this respect Kenya seems to have struck a fortunate balance. There is still a strong body of European officers, particularly in the technical field, but it can definitely be said that the Kenya Africans are masters in their own house — at least as far as agricultural policy and administration is concerned. There has undoubtedly been more Africanisation in agriculture than in trade or industry.

II. African Socialism

In Kenya Africanisation and Socialism seem to be closely linked. The term "socialism" is applied not only to a change in the economic or social field, but also to measures which promote a racial change without touching the class structure. In its statement in May 1965, the Government of Kenya clarified what was officially understood as "African Socialism" and this statement was unanimously approved by the House of Representatives, i.e. the political body. The points relevant to agriculture are the following:

— (24) Idle resources and undeveloped land will be taxed in ways designed to encourage efficient use and development.
— (26) Emphasis will be given to the development of agriculture in former African areas through land consolidation, registration of titles, development loans, co-operatives, and extension services.
— (27) A working party will be established immediately to consider and recommend on forms of land tenure throughout the country.
— (28) Land management legislation, including punitive measures against those who mismanage farms, misuse loans, default on loans, refuse to join major

[1] In Kenya the principle that administrators are not allowed to interfere in politics has been kept, while in Tanzania politicians have been nominated as administrative heads of Regions and Districts.

co-operative farming schemes where these are necessary, or oppose land consolidation, will be introduced and strictly enforced.

— (29) Agricultural land will not be sold to non-citizens unless by consent of the Government.

— (30) The organisation and functions of marketing boards will be re-examined with a view to consolidating their activities and modifying their functions to promote the welfare of consumers as well as producers.

— (31) Priority will be given to producer co-operatives in making future agricultural development loans.

KANU's African Socialism does not suggest the immediate nationalisation of large farms owned by Europeans and Asians, something which would not be unusual in a country where the landed properties of minorities are as conspicuous as in Kenya. KANU is propagating an evolutionary approach to the problem of Africanisation and Socialism. Kenya is thus following the pattern of almost all newly independent countries in Africa south of the Sahara, which is different from the procedure in Egypt, Algeria and Tunisia where large European farms were nationalised with hardly any compensation whatsoever. The difference in policies is clearly the result of a difference in circumstances. The fact that Kenya's economy is highly vulnerable to any drop in agricultural production, the lack of African personnel who could run large farms, the dependence on aid, particularly from Britain, and, not least, the fact that almost every politician, official and functionary has recently acquired some property, is acting as a powerful brake on any suggestion of immediate nationalisation. "It would be clearly illogical", said B. McKenzie in the House of Representatives, "to deplete our small capital resources by nationalizing the agricultural industry... It would be far more advantageous for us to utilize our ... resources for new agricultural development rather than a mere transfer in ownership ... our emphasis must swing from (re)settlement without any significant economic growth to an intensification of efforts in underdeveloped areas" (May 5th, 1965).

Of particular interest is the emphasis on the co-operatives in order to establish an identity between those involved in production and those receiving the fruits of that labour. The British administration, which gave strong support to co-operative trade, had little faith in co-operative land use, remembering the failure of group farms in Nyanza and of grazing schemes in semi-arid areas. The African Government is taking a different attitude. Some politicians are telling the peasants that small pieces of land cannot play a part in the agricultural industry of Kenya and that they should pool their land and work as a co-operative. The Minister for Agriculture, B. McKenzie, expounded in the House of Representatives: "... we will continue to give generous financial support to the Land Bank in order to facilitate the acquisition of land by our people ..." but "... we would prefer to see the larger farms passing into the hands of competent, well-organized, properly-run and self-disciplinized producers' co-operative societies . . . particularly by the workers already employed on the farms . . ." (May 5th, 1965). However, the actual steps taken in this direction indicate that the Kenya Government prefers a much more cautious approach than does neighbouring Tanzania. All types of landownership: state farms (national farms), co-operatives, companies, partnerships and private holdings are acceptable. National farms and co-operative enterprises are, as a rule, not considered as alternatives to smallholder farming but as supplementary, according to the circumstances.

There is no administrative or political pressure in favour of co-operative land use, but incentives are provided by specific aspects of the loan policy. Maize growers who pool their land into blocks of a minimum size of 10 acres are eligible to apply for a minimum return of shs 120 per acre [1]. A number of co-operative farms came into existence in the Highlands as a result of generous loans given to former farm workers. Some dry land south of Machakos, bought by the "Settlement Fund Trustees", has been put into co-operative ranches, the institutional arrangement of which seems to be much more like a company than a co-operative. Land in the Ol-Kalou salient which is unsuitable for distribution to smallholders is planned to go into a few large co-operative farms. But there is no popular upsurge for communal and co-operative land use in Kenya as there was in Tanzania a few years ago.

It is interesting to note that efforts in communal farming are made in those parts of Kenya where the conditions of production are similar to those in Tanzania. Cultivation in communal farms spontaneously came to existence on the coast of Kenya as well as of Tanzania, but rarely in the hinterland of either country. The block-cultivation of cotton, with everybody having a piece of land in the block, has not only been established in Tanzania's Sukumaland, but also in the Nyanza Province of Kenya, where it is called a "Saga block". According to Omamo, two types of group effort can be distinguished in Nyanza. (1) A man calls upon his relatives and friends in the vicinity to assist him with the field operations. In return he prepares plenty of food and serves native beer. (2) A clan organizes itself and works as a team on each field in due turn. All that the member is required to do is to provide some light refreshment.

This second type has become more popular and is being encouraged. Early in 1964 there were approximately 500 Saga groups having planted 4,000 acres of cotton in blocks ranging from 3—30 acres each. The leader of the group is usually a respected clan elder. Individual rights are retained in most cases and each family works on its own piece within the cotton block. The Government provides a loan for contract ploughing and spraying (usually by air). Mechanised ploughing and

[1] The system of a "minium return" was introduced in the large farming areas in order to reduce the risks of production. A farmer who obtains a "planting order" from the local Agricultural Committee, a body of experienced farmers of the area, is liable to receive a loan for working capital which need not be repaid if the Agricultural Committee acknowledges that there is a crop failure which is not due to poor husbandry.

spraying means high costs per acre of cotton, which can only be met if yields of 700—900 lbs per acre are reached. It is open to some doubt whether the Saga blocks will remain popular, when the tight economics involved are realised by the peasants concerned [1].

Much more important than the emphasis on co-operative land use is the support for land consolidation and registration. The Government and the political party, KANU, are apparently convinced that individual ownership of land provides a greater stimulus for agricultural production than does collective or communal ownership. The KANU manifesto states: "Every farmer must be sure of his land rights." Even those who are usually considered the spokesmen for the left wing within KANU, support land tenure reform in the former Reserves leading to "absolute ownership" of land in private hands. It is particularly encouraging to see that land consolidation, enclosures and farm layouts continue to make progress in Nyanza where political opposition was experienced in the past.

Kenya's African Socialism is thus actively promoting the small agricultural entrepreneur, the man who owns somewhat more land than is needed for subsistence, who employs seasonal labour, who is perhaps more a foreman than a labourer himself, and who has, during the last few years, been the driving force of agricultural progress in the former Reserves. African Socialism also supports the resettlement programmes, which include the establishment of yeoman farmers; and it does at least accept the present support for African large farmers. T. Mboya, Minister for Planning, says in this respect: "The urgent need to develop the agricultural sector should not obscure the equally important need for land tenure reform to ensure that this development is not concentrated in the hands of the few. On the other hand we need to bear in mind that we cannot limit individual ownership only in agriculture. Any limitation of individual ownership has to cover all sectors. It is therefore not something to be done haphazardly, and it is for this reason that the Government has suggested that a working party might be established to consider and recommend on forms of land tenure and possible limitations on property, and to advise on the machinery for making these effective."

The Government's interpretation of African Socialism is not, of course, accepted by all. It is contended that African Socialism is little more than an umbrella for the rapid spread of petty capitalism, that it is an instrument for the establishment of a "black landed gentry", that private

---

[1] The price for disc-ploughing of cleared, arable land in Kenya varies between 40—50 shs per acre where a tractor is used. For contract ploughing with oxen 15—25 shs are charged in Nyanza Province. When comparing these figures, it must be taken in account that ox-ploughing is rarely done in time and that the quality of the ploughing is much poorer, particular on heavy soils.

ownership and exploitation are two inseparable aspects of life, that the Government should try to revitalize the communal life of the old tribal times which resulted in "a fair distribution of wealth and intensive co-operative or team spirit." The land problem has been the backbone of the political struggle and is still the main topic of political discussion. B. KAGGIA, a major spokesman of the opposition group within KANU, argues not so much against smallholders in the former Reserves as against African large farms. "For years and years", said KAGGIA, "Africans had complained about the European settlers owning miles and miles of land. It would be ridiculous to allow rich Africans and Asians to take their place. If we had 20,000 rich Africans taking over all the land that was formerly owned by Europeans, then we shall have achieved nothing, we shall have only exchanged white settler exploiters by African settler exploiters" (7th May, 1965). Other backbenchers are reflecting the feeling of a sizeable minority in the former Reserves by arguing that "most of the ministers here ... might have been involved in this buying of large areas of land ..." while thousands of landless "are crying for land, for employment, for a place under the sun" (22nd April, 1965). Another major bone of contention is the fact that settlers in resettlement schemes have to pay two-thirds of the purchase price of the land. "We fought", said M. ANYIENI, "so that we could have our land back ... The Africans now are forced to pay for it ... there is no reason why, after you fought somebody who had taken your property, you have defeated him, that you should have to pay him".

Although several aspects of Kenya's African Socialism are still being disputed, it can certainly be said that, together with political independence and Africanisation, it has contributed to the establishment of some degree of identity between the people of Kenya and their Government. Experience in East Africa indicates that political independence has considerable economic advantages for Africans who have a reasonable level of education, even though this might not show up in the estimates of the gross domestic product. Salaries in Government and industry which used to go to Europeans and Asians are being paid to Africans. More aid is forthcoming, partly from the former colonial power and partly from other sources. There is more hope, more confidence, and more social mobility, which has concrete repercussions in the form of additional economic activity. Perhaps some services no longer run as smoothly as they used to do, but the increase in the general activity offsets the economic losses of Africanisation. In rural areas the people have become more open to technical progress. It is easier to introduce innovations. The political change has, perhaps, done more for the spread of cotton cultivation or the use of mineral fertilizer than did a decade of well run British extension work. The political party, KANU, is no longer acting as a brake on agricultural development, which is particularly important in Nyanza where the opposition to the agricultural policy of the British was the fiercest. The political fever that affected some districts in 1963 has gradually disappeared. Many more Africans than before are busy "making money" for themselves, and only a few are busy encouraging peasants to undertake communal tasks. There seems to be far less enthusiasm and far less illusion about Independence in Kenya than there was, for instance, in Tanzania, but much more African economic activity is going on than in British times, and this might prove to be a sounder approach to economic development and to the attainment of full economic independence.

On the other hand, it should not be overlooked that Kenya has to forego some of the push to agricultural progress which is possible if a unified political party is actively supporting the efforts of the Agricultural, Veterinary and Settlement officers. Most of the local politicians in Kenya seem to be preoccupied with the defence of short-term tribal interests which more often than not are detrimental to the development of the nation and thus in the long run to the interests of the people they represent. It is most unfortunate that KANU is split into factions with different views as to agricultural policies and different attitudes towards civil servants. The Government will have to clamp down on loan defaulters in resettlement schemes and in the former Reserves, but this is a highly unpopular move which can succeed only if the local politicians are fully behind the policy of their government. There is, and here we support M. Kibaki, the "urgent need to revitalize KANU", to bring its factions together and to use KANU as an instrument for rural discipline and progress (East African Standard, July, 1965).

III. The Development Plan 1965—1970

The aspirations towards Africanisation, African Socialism and economic development have been crystallized in the Five Year Development Plan for 1965—1970. The major objectives of this plan as regards agriculure are:

- to accelerate the process of land consolidation and registration. The estimated expenditure for this purpose amounts to more than £ 2 million (See Table 26);
- to provide the resources for a direct stimulus to increased production, particularly in the former Reserves. Most of the estimated £ 4 million for general agriculture, £ 6.6 million for agricultural credit, £ 2 million for livestock development and £ 1.8 million for land reclamation serves this purpose;
- to stabilize and expand the economy of the former Scheduled areas. This is by far the biggest item, comprising not only the £ 13 million earmarked for resettlement but also the £ 6 million for the Land Bank;
- to develop irrigation farming on substantial acreages at an estimated cost of £ 7.6 million.

The production targets are laid down in Table 27. A 70 per cent increase in marketed production is expected. The expansion of the coffee, tea and sisal production is planned to carry most of the increase. With coffee,

Table 26. *Estimated expenditure for agriculture, livestock, land consolidation and settlement in the Five Year Plan, 1965—1970*

	£	in %
Agricultural Credit including Land Bank	12,200,000	29
General Agriculture, Research, Education, Extension and Tea Development	4,006,400	9
Land Reclamation, semi-arid areas mainly	1,827,000	4
Livestock	2,006,100	5
Irrigation	7,617,000	18
Land Consolidation	2,053,000	5
Land Settlement	13,084,000	30
Total	42,793,000	100

Source: Development Plan, 1965—1970; p. 65—71.

Table 27. *Estimates of marketed agricultural and livestock produce in 1970, in £'000, return to producers*

	1954	1962	1970	1970 as per cent of 1962
Coffee	5,460	9,480	16,870	179
Tea	2,060	6,690	10,920	163
Sisal	2,240	4,520	8,500	188
Cattle and Calves for slaughter	2,200	4,790	5,150	108
Dairy Products	3,910	5,300	6,920	130
Sugar	280	1,170	3,970	339
Cotton	790	460	4,460	1,000
Maize	5,970	3,330	3,900	117
Pyrethrum	620	2,210	3,000	136
Wheat	3,860	2,140	2,710	127
Wool	220	370	1,480	250
Rice	—	310	1,110	358
Other	5,700	5,980	10,260	173
Total	33,310	46,760	79,250	170

Source: Development Plan 1965—1970; p. 49, Table 1.

the additional sales will not come from additional plantings but mainly from recent plantings which will reach maturity during the Plan period. Particularly striking is the Plan target for cotton, which is ten times the present production. Rice and sugar are two other products which are expected to show up with significant increases.

It is obvious to any observer of Kenya's agriculture that the 1965—1970 Development Plan is unlikely to become implemented in every detail. It should be considered as a guide line for Government policies in a difficult transitional period. Nevertheless, the Kenya Plan is much more realistic than the extremely ambitious one published in Tanzania in 1964.

E. Future Aspects of Agricultural Policy

I. Indications for a New Appraisal

The final ten years of British agricultural policy in Kenya — as far as it concerned the Africans — showed a number of features typical of the British Colonial administration in many parts of Africa, and which involved more funds, personnel and efforts in Kenya than anywhere else. These are as follows:

- The main starting point for economic development was seen in the promotion of smallholder agricultural production. Only a minor proportion of the public funds which Britain made available for Kenya went into the promotion of non-agricultural industries.
- The instrument for the promotion of the African smallholder was an unusually close and well organised network of agricultural development services run by the relevant ministries. Within these services almost all positions of importance were occupied by British Officers with a relatively high degree of technical and economic experience, but living in a different world and having a rather different set of values from that of the Africans.
- The approach to the smallholder, strongly influenced by the failure of direct compulsion during the campaign for soil erosion control in the early fifties, was based on the persistent persuasion of the peasant to grow cash crops. This was combined with the close supervision of major activities by a large number of minor employees. With the approach of Independence, the element of pressure still contained in "close supervision" (for instance of coffee husbandry) more or less waned, leaving merely the appeal to jump on the bandwagon of the "money-makers" which was set in motion by services like consolidation, enclosures, extensions, loans settlement, markets and processing.

Considering the increase in marketed production, the standard of husbandry, and the great number of innovations firmly adopted by the smallholder, it can certainly be said that no other colonial administration achieved so much in so short a time. Smallholder development in Kenya between 1955 and 1964 is probably the best that can be shown by any colonial administration which is not in a position of undisputed authority and which does not want to use outright compulsion.

However, a number of problems of growing importance which result from the approach of the last 15 years should not be overlooked:

- A heavy administrative apparatus, with personnel in almost every village or location and with many unusually highly paid officers, has been built up. Too many personnel are grossly overpaid in relation to the country's situation.
- The expenditure for agriculture has been rising much more rapidly than has the value of the agricultural production and this puts an increasing burden on the Treasury. Kenya's civil servants are probably among the best in Africa south of the Sahara. However, Government departments everywhere in the world

(1) are top heavy, (2) shy away from quick or unpleasant decisions and (3) are not in a position to dismiss unsuitable personnel. Wherever possible they should delegate the work to more appropriate bodies.

Table 28. *Expenditure for agricultural services and investments by the Governments of Kenya, Tanzania and Uganda in relation to the value of total agricultural production* [1, 2, 3]

	Kenya			Uganda			Tanzania		
	Expenditure	Production	Expenditure as a % of production	Expenditure	Production	Expenditure as a % of production	Expenditure	Production	Expenditure as a % of production
	£m	£m	%	£m	£m	%	£m	£m	%
1952	4.1	.	.	1.8	.	.	1.4	.	.
1953	4.0	.	.	1.3	.	.	1.5	.	.
1954	6.0	74	8	1.0	86	1	1.5	81	2
1955	6.6	75	9	1.0	90	1	1.4	92	2
1956	5.6	83	7	1.5	87	2	2.0	95	2
1957	4.8	85	6	1.4	90	2	2.1	99	2
1958	5.8	87	7	1.4	87	2	2.3	98	2
1959	5.7	89	6	1.6	87	2	2.2	105	2
1960	6.8	90	8	1.7	86	2	2.5	109	2
1961	7.8	87	9	1.5	88	2	3.6	106	3
1962	11.4	103	11	1.4	87	2	3.7	118	3
1963	15.5	111	14	1.4	98	1	3.9	120	3

[1] Recurrent and capital expenditure for Agriculture and Veterinary.
[2] Monetary and non-monetary production as estimated in the GDP.
[3] The figures for expenditure concern the fiscal year from 1st July to 30th June. The figures for production concern the year from 1st January to 31st December.

Source: Government of Kenya: Statistical Abstracts. Economics and Statistics Division, Nairobi.

— The increase in services is sapping the country's power of internal capital formation. The expenditure for agriculture has more than trebled over the last 12 years. The rate of increase in agricultural production is not the same, indicating declining marginal returns for additional public expenditure. Consequently, a high and growing share of the value of the total agricultural production — rising from 6 per cent in 1957 to 14 per cent in 1963 — has been consumed by Government expenditure related to agriculture. The situation in Tanzania and Uganda is different. The percentage of the total production which is consumed by Government expenditure is far smaller, ranging between 1 and 3 per cent. Naturally it would be unfair to compare the relevant figures of Kenya, Tanzania and Uganda without mentioning (1) that a significant proportion of Kenya's expenditure is caused by the need to africanize the Highlands and (2) that substantial funds are used for long-range investments: land tenure reform, tea development, sugar development, irrigation etc. However, the discrepancy between public expenditure for agriculture and the value of additional agricultural production is so great that there is reason to doubt whether it can be sustained for very long.

— The new African administrators and their ministers inherited from their colonial predecessors — who were administrators by profession and some even by vocation — the idea that administrations are omnipotent. As soon as there are shortcomings, there is an ouctry for more personnel in the relevant department and for more finance to pay and equip this personnel.
— It is only natural that British officers, searching for funds to promote smallholder development, looked to Britain for additional funds. This "colonial" attitude, however, is lingering on. The observer of the Kenya scene in 1965 has the impression that there is still too much reliance on aid or loans from abroad. Admittedly, the streamlining of existing services, schemes or farms is a most unpopular, tedious and unrewarding job, which requires above all political backing [1]. But doing the job is a prerequisite for economic independence.

Great as the success in the important sphere of smallholder development might have been, the costs have been heavy and have been carried largely by the British tax payer. The basic element of economic development in Kenya has actually been the tutoring and subsidizing of agriculture by British funds and British personnel. The policy in the last decade of British rule was perhaps the best possible under the given circumstances. However, circumstances have changed. Kenya cannot continue to live indefinitely on foreign subsidies for her agriculture. If Independence is to mean something, the burden of development must be placed on the people of the country. The smallholder must be considered a mature commercial producer — although this is not always the case — because if he is not, it will perhaps take too long a time for the peasants to adapt themselves to the demands of a modern economy.

Several questions arise out of all this: What changes are pertinent in the agricultural development policy? What changes are already planned for or discussed? What is their chance of success? It is beyond the scope of this report to deal with these questions in detail. It is the purpose of the following pages to inform about important trends in production development policies, as they were recognizable in mid-1965, and to discuss briefly the problems connected with them.

II. The Importance of Urban Purchasing Power for Progress in Agriculture

One of the first points which should be kept in mind is the importance of urban purchasing power to the future of smallholder farming in parti-

[1] The reliance on aid is not only understandable, but also economic from the standpoint of a man in charge of some institution. Let us assume that there is an agricultural school with a large farm attached to it. The principal may have the choice of using his limited time in order (1) to increase farm profits with which to pay for the school or (2) to go to several aid agencies and ask for support. It is most likely that the second approach will yield much higher returns.

cular. The prerequisite for any farmer to produce is the existence of buyers at price levels which are conducive for the intensification of land use. In Kenya the number and the purchasing power of the buyers of agricultural produce is rather limited. Everyone who attaches priority to agriculture — as was done under the British administration — at the same time assumes that the produce will be exported and that there is an attractive market for it. In the mid-fifties, when British efforts gained momentum, the prices for several export products, in particular for coffee, were extremely good and the priority attached to agriculture was undoubtedly justified. However, the market situation has changed:

— The crop which has played the leading role in smallholder development, coffee, is bought almost exclusively by consumers in Central Europe who are supposed to have a moderate income elasticity of demand. This means that a rapid increase in production is only likely to produce falling prices. Furthermore, the rulings of the International Coffee Agreement have to be obeyed. Additional plantings of coffee are prohibited, and this has, understandably, caused widespread discontent. Smallholder development in Kenya would be bright and easy indeed if more coffee could be planted. No other event has put such a strain on future African agricultural development as the restrictions on coffee. For other crops the market forecast is perhaps not as poor, but they are either expensive to establish (as, for instance, tea) or the economic impact is far smaller. On the whole the emphasis on the export market is not likely to offer sufficiently large and attractive prospects for sustained smallholder development.

— The income expectations [1] of smallholders and the wages for rural labour in Kenya are probably lower than in many South-American, Middle-American or Asian countries, the exports of which compete with Kenya's. However, the performance per man hour of actual work is probably significantly lower. Consequently, the wage costs (or income expectations of family labour) per unit of output can be assumed to be higher than in many, if not most, countries with competing exports. Thus, in the main, the only agricultural activities in Kenya which are likely to be competitive with the produce from elsewhere in the world are those which fit in particularly well with Kenya's soil and climate — for instance, coffee, tea, pyrethrum, sisal, etc. For these products the location actually provides the producer with a rental income.

The figures of Table 29 seem to confirm the above hypothesis. A comparison of the estimated labour inputs per acre, per lb. and per 100 shillings production value of coffee in smallholdings in Nyeri District, large farms in Brazil, smallholdings in Columbia, large and small farms in El Salvador, indicates that the man-hour requirements on Kenya's smallholdings are much higher than elsewhere. The labour performance in Kenya is apparently poorer than in Middle or South America.

This is probably true not only for coffee but for agricultural activities in general. With the restrictions imposed on coffee, the limited quota for

[1] The level of income, per hour of work in smallholdings at which peasants are willing to put significantly more effort into farming.

pyrethrum, the unavoidable slowness of tea development and the ecological limitations of cotton, smallholder expansion in Kenya depends mainly on the market for milk, maize, vegetables, rice, starchy roots, bananas,

Table 29. *Labour input in coffee production on smallholdings in Kenya and in several American countries*

	Man Hours per Acre of Mature Coffee	Yield in lbs[1] per Acre of Mature Coffee	Man Hours per lb[1] Coffee	Price in Shs/lbs[2,1] Coffee	Man Hours per 100 Shs Worth of Produce
Kenya, smallholdings[3] Nyeri District	1933	1360	1.4	2.70	53
El Salvador, small and large holdings	630	590	1.1	3.05	35
Columbia, smallholdings	323	465	.7	2.70	26
Brazil, large holdings	233	395	.6	1.40	42

[1] Dry parchment.
[2] Price to producer.
[3] Five year old coffee only. The data for the other countries concern mature coffee of different ages.

Source: 1. Farm Economic Survey: Some economic aspects of agricultural development in Nyeri District. Rep. No. 21, Nakuru, 1964.
2. FAO: Coffee in Latin America. Vo. I. Columbia and El Salvador, New York, 1958, Vol. II. Brazil, State of Sao Paulo, New York 1960, Part 1.

poultry, pigs, pulses etc., that is on types of goods whose sales and prices are largely determined by the purchasing power of the internal market.

The need for internal purchasing power coincides with the need for employment of both the urban and rural population. Kenya's population is growing at a rate of about 3 per cent a year. At this pace there will be 15 million people in Kenya by 1980. The idea that the unemployed should go back to the land is as old as the history of urban communities. It is valuable only as a temporary measure. Sustained efforts in this direction failed in Roman times and ever since, whenever they have been attempted. In Kenya unemployment is no longer a relative term as it is, for instance, in Uganda which still has vast stretches of fertile soil lying uncultivated. The unemployed — in the main — are no longer those who shy away from peasant farming in order to get an "easier" job in town, but are more or less rural surplus population. Unless adequate steps are taken to limit the number of human beings attempting to live in any area, the best efforts of the Government to develop agriculture will be futile. Economic history shows that there is only one answer to urban unemployment and rural underemployment: urban employment. Thus there is urgent need in Kenya for productive urban investment in order to (1) absorb the urban unemployed and (2) provide them with the purchasing power which is the prerequisite for agricultural development.

The demand for productive non-agricultural investment poses a number of questions among which the shortage of funds is crucial. It is obvious that development aid can only fill part of the gap. Historical evidence and model calculations show that there is normally only one major source for the financing of capital formation in an emerging economy: the profits and revenues from primary production, i.e. from mining and agriculture. In the case of Kenya we may add tourism. Since mining is of minor importance in East Africa and since the weight of tourism is certainly not sufficient to carry most of the burden, no other choice remains but to "milk" agriculture in order to obtain the funds for financing urban development, as was done in Germany, Russia, Japan and almost everywhere where sustained economic development took place [1]. This again implies that there is little room for subsidizing agricultural schemes, as was widely done by the British; public measures for agricultural development must be profitable in terms of the Treasury. The establishment of nice looking technically successful schemes which are heavily subsidized either by granting generous conditions to the participants or by charging the costs of personnel to some Government department may well be worthwhile in industrialized countries, but simply does not fit into Kenya's present situation.

In Kenya, as everywhere in emerging economies, it is possible to "milk" agriculture in order to "feed" non-agricultural growth without using direct compulsion as was the case in Russia's early development, and without the hardships which the landlords in Japan imposed on their tenants one hundred years ago. The process of squeezing agriculture, in particular smallholder agriculture, can even be accompanied by growing per capita incomes in rural areas. The very simple and straightforward answer is: better husbandry. Good husbandry, to a certain degree, is also the answer to crop failures in years with marginal rainfall. If established principles such as terracing, tie-ridging, early planting, early weeding, use of suitable varieties and choice of suitable crops were applied, they could significantly reduce the danger of famines. Another approach is the promotion of cash crops. Farmers with well husbanded cash crops, even if the maize fails, are in a position to buy foodstuffs out of their cash income.

---

[1] See: Johnston, B. F., and Mellor, J. W.: Agriculture in Economic Development. American Econ. Rev. Vol. II, No. 4. Sept. 1961. — Leibenstein, H.: Economic Backwardness and Economic Growth, New York 1957, p. 54. — Lewis, W. A.: Die Theorie des wirtschaftlichen Wachstums. Tübingen-Zürich, 1956, p. 243. — Nurkse, R.: Problems of Capital Formation in Underdeveloped Countries, Oxford 1955. — Ruthenberg, H.: Landwirtschaft und anfängliche Kapitalbildung. S. H. 2. Zeitschrift f. ausl. Landw. Frankfurt 1964. — Rostow, W. W.: The Take-off in Self-Sustained Growth. The Econ. Journal, March 1956.

It is obvious to every agriculturalist that the production potential of Kenya's land in the small as well as the large farming areas has, as yet, hardly been touched. It is certainly most unfortunate that the education of Kenya's agriculturalists is largely influenced by the pattern of land use in England, the U.S.A., Russia, and Israel, i.e. by the performance of countries whose agricultural development has little relevance to the likely expansion path of most of Kenya's smallholder farming. The farmers of Central Province, for instance, could benefit immensely from the experience accumulated in the Highlands of Asia, particularly in Japan or Taiwan. Much, if not most, of the land is undeveloped or underdeveloped compared with standards in Japan, Taiwan, India or Italy. Rural underemployment goes side by side with low intensity of land use and large uncultivated areas, even in densely populated Central Province [1].

More important than the increase in acreages are the prospects of higher yields per acre by proper planting, manuring, weeding, spraying, pruning and harvesting. The same holds true of increasing livestock output per acre by proper raising, feeding, dipping, serving of cattle, sheep, pigs and poultry. Some finance is, of course, involved in good husbandry (fertilizer, insecticides, seasonal labour etc.), but the return comes back within a few months. Comparatively few capital investments are required and most of them can actually be financed out of profits from good husbandry. Centrally organized services — extension, seed supply, credit, markets and processing — are already available in Kenya and are much better organized than almost anywhere else in the tropics. Thus the additional costs of increased smallholder production could be negligible. It is crucial for economic development that (1) the production potential, which can be developed mainly by good husbandry, is realized, that (2) a significant part of it is channelled as revenue into the Treasury's coffers, and that (3) the Treasury does not waste it on conspicuous Government expenditure, in particular on salaries which are out of proportion to the country's financial position, but invests it in productive enterprises, which in due course provide purchasing power and thus markets for agricultural produce at prices which are an incentive to further production.

[1] It is interesting to observe that most of the valley bottoms even in densely populated areas are used for extensive grazing only. Here and there hesitating efforts are made to reduce water logging by digging ditches. Hardly any efforts are made to control and use the water for small-scale irrigation. In this respect Kenya could certainly learn from several centuries of experience in Madagascar.

III. Future Aspects of Smallholder Development in the Former Reserves

Smallholder farming in the former Reserves is the most important form of economic activity in Kenya, both in terms of numbers employed and the contribution to the Gross Domestic Product. The Kenya Government realizes its importance and has stated repeatedly that in the future allocation of resources priority will be given to development in the former Reserves. The necessary experience is available; the research services are far in advance of the application of tested innovations; the task ahead is the wide-spread application of good farming practices.

Kenya is one of the few tropical countries with relatively high rates of increase of production in smallholder farming, which indicates that the process of mobilising the smallholders' resources is well under way. Crop and animal husbandry is still poor compared with that in most Asian, Mediterranean or European countries, but much better than that in most smallholder farming areas in Africa south of the Sahara. A remarkable proportion of the farmers now understands the advantages of cash activities, better seeds, row cultivation, manuring, fertilizing, the dipping of cattle and artificial insemination. The most important aspect might be the general acceptance of mineral fertilizer together with early weeding. Whereas fertilizers were used previously on a small scale in coffee only, they are now used increasingly on such crops as tea, maize, potatoes and even on leys.

1. Possibilities and Limitations of "Close Supervision"

It is, however, obvious to the observer in the smallholding areas that the very instrument which was successfully used by the Agricultural Department — viz. close supervision of important cash activities by numerous inspectors and assistants — has become questionable. In the early 1950's, when cash activities were still small, this approach worked amazingly well and established a pattern of husbandry which was of a very high standard. But with the growing scope of commercial activities the "close supervision" approach shows major shortcomings:

- It is becoming more and more difficult to keep control of husbandry. It is certainly possible to control 1000 acres of coffee spread over 1000 holdings with the agricultural staff usually available at the District level, but it is impossible to control 15,000 acres of coffee, 2,000 acres of pyrethrum and 10,000 grade cows.
- If complete control of all plantings, all grade cows and all loans is desired, the staff numbers must increase at almost the same rate as do smallholder marketed sales. This would put an insupportable burden on the Treasury. Perhaps even more important is the argument that like almost all newly independent countries, Kenya is experiencing difficulties in having an efficient civil service. A rapid increase of minor field staff is likely to create insurmountable problems of supervision and control.

Owing to these two factors, a general policy of close supervision of the husbandry of all cash activities, as it was exercised in the mid-1950's, can no longer be advocated. The economics of staff-use leave hardly any choice but to restrict the "close supervision" approach in the former Reserves to (1) the initial stages of important innovations and (2) crops of special importance like tea, the supervision costs of which are carried by a special cess.

In order to relieve the financial burden on the Treasury, the task of close supervision of coffee husbandry in Central Province was transferred to the coffee-co-operative societies in 1964. The minor field staff responsible for coffee were henceforth employed by the societies and paid for by their cess on coffee. This seems to have made the situation even worse. According to the Annual District Reports, the state of coffee husbandry leaves much to be desired and this is generally attributable to the lack of stern measures on the part of the society committees against negligent farmers. It is doubtful whether pressure on the co-operatives will yield lasting improvements in husbandry. Experience in Kenya and elsewhere shows that supervision must be imposed by personnel who are answerable to some outside body, with the headquarters in the capital town, far away from the district, which is not involved in the social and political affairs of the locality.

Probably no choice is left but to accept that the husbandry standards — for instance in coffee — which have been artificially high, considering the traditional peasant setting, will in future deteriorate; that there will be a growing differentiation between good and poor farmers; and that there is hardly any economically worthwhile direct way of persuading the negligent farmer to improve his standards. Government's agricultural personnel cannot be the friendly tutors of everyone who is not looking after his own affairs properly.

2. Possibilities and Limitations of the "Voluntary Approach"

Since the "close supervision" method has withered away to something of little more than nominal value — with the notable exception of tea, which is supervised by the KTDA — smallholder development relies almost exclusively on the incentive to earn cash. The "voluntary" approach has undoubtedly been a success. The relation between the peasants and the agricultural staff, which was rather strained in the early 1950's, was excellent in the mid-1960's. The agricultural development services of the Government are in higher demand than they have ever been. The Farmer's Training Centres prove to be the mainstay of agricultural education and extension. Good progress has been made with agricultural youth clubs ("4 K" Clubs), which have a great appeal [1].

[1] In mid-1964 Kenya had 305 local 4-K Clubs with a membership of 8,000 youths. The "4 K's" are: Kuungana (to come together), Kufanya (to make), Kusaidia (to help), Kenya. They are mainly concerned with vegetable growing and, occasionally, with raising rabbits.

However, promising as the situation might be, it should not be overlooked that the "voluntary" approach has serious limitations. Progress has generally been limited to:

— rather advanced areas. This is particularly true for Central Province where a lot of pressure had been exerted before, during and after the Emergency. It might well be that the "voluntary" approach works because a decade of pressure has laid certain foundations;
— somewhat larger farmers within a small farming community. The better farmer is usually the one who not only works longer but who has some cash available to employ labour and therefore works longer hours as a foreman supervising his labour rather than as a labourer himself. Rural progress has apparently been made not so much by the hard-working self-contained family unit, but by those who recognize the possibilities of cash production and employ people who do not work for themselves, either because they do not see their chances or because they lack the necessary land.
— activities which yield an unusually high return per hour of work. Take away coffee and its secondary effects from Kenya's rural scene, and there is not much left. Coffee, however, is easy money.

The majority of the smallholders in the former Reserves are still sitting on enormous capacities of semi-idle labour, land and cattle and make only slow progress. There is tremendous diversity within the smallholder community. The performance is by no means uniformly good or bad according to the quality of soil and climate, but almost everywhere there are some good farmers reaching a level of husbandry which compares favourably with good farming elsewhere in the world, side by side with large acreages which are negligently used. On the other hand, there are areas where the rate of development in improved agriculture is alarmingly low, where there are few signs of any determined effort to overcome the problems and where much of the work done under British pressure is falling into decay. Most peasants are still inhibited by the idea that their soil is no good, that the climate is adverse, that increased efforts yield but low returns and are not worthwhile. Very few realize the importance of putting much work into something which has slow but accumulating results. Most farmers feel they are not able to increase productivity by their own efforts without some outside assistance. The fallacious idea is prevalent that there must first be additional finance for seasonal labour, tools, stock, seed and fertilizer, while in reality the first step is the improvement of husbandry: proper planting, weeding, manuring and harvesting. There is a tendency — perhaps a colonial heritage — among a large section of the smallholder population to depend almost entirely on the Government for the solution to their problems, to wait for grants or loans, instead of trying to help themselves. Land consolidation or enclosures are not enough to mobilize a dormant peasantry. Strong additional pressures, incentives or stimuli — as experienced by Central Province during and after the Emergency — are needed if the consolidated land is not to go back to primitive subsistence farming, extensive pasture or even bush.

It is frequently argued that more staff and more loans are required to overcome the lethargy of that section of the smallholders who have not yet been drawn into the exchange economy. Spencer, for instance, in his 1964 report for Elgeyo-Marakwet writes: "I must stress the highest priority in farm extension staff, who can teach the people the necessity to work and to obtain money to raise their standards of living ... It is disappointing to see the potential not being exploited due to the lack of staff and poor co-operatives". However, the number of additional staff that can profitably be employed is limited. There is something like an "optimum" in the use of agricultural staff. In some districts of Kenya we might expect rapidly decreasing marginal returns if any more staff are employed. The real problem is that of (1) minor staff not being used effectively enough and (2) the majority of the smallholders being slow to accept the innovations propagated by the staff. Considering the great number of staff already employed, the financial stringencies of the Treasury, and the need for priority investment in industrial development, it seems that the intensification of the "voluntary approach" by more staff might not be sufficient to mobilize the production potential of smallholders rapidly enough. The "voluntary" approach must probably be supplemented by some form of "pressure" [1].

3. Possibilities and Limitations of "Direct Pressure"

In Tanzania efforts are made to solve by political pressure the problem of those large masses of smallholders who do not react satisfactorily to incentives. Village Committees have been established and they are expected to set cultivation targets of several acres of cash and subsistence crops per family. The TANU secretary and the TANU Youth League are supposed to induce the peasants to toe the line. In Kenya there is no such pressure and in most areas it would have no chance whatsoever of succeed-

[1] The success of extension depends largely on drive, sound economic judgement and long local experience by the man on the spot. It is certainly not by chance that the idea of agricultural extension as being something relying entirely on the "voluntary approach" came from highly industrialized countries, was carried into Africa by personnel from these countries and has worked properly only under the umbrella of colonial government or an administration schooled by colonial officers. This might be considered an indication that a certain stage of social mobilisation and long experience with the demands of the industrialized society are essential for the " voluntary approach" to work effectively. It is therefore open to some doubt whether agricultural extension of the kind introduced by the British will last long after Independence. It can be assumed that the newly promoted African Agricultural Officer will much prefer to be assigned clearcut administrative duties which he can execute than to have the task of selecting local priorities, developing innovations on his own and persuading farmers to follow them.

ing. In many parts of Tanzania there are still large stretches of uncultivated land which — although economically marginal — can be cultivated. Under such conditions it is possible to increase the acreage and perhaps the production by political pressure. In the Nyanza Province of Kenya the expansion of the cotton acreage might offer some opportunity for a similar approach. As a rule, however, Kenya's agricultural development cannot depend on increased acreages of certain crops but on better husbandry, and it seems most unlikely that better husbandry can be achieved with the help of any "party whip".

Harsher measures are needed on certain specific items which are covered by administrative ordinances: erosion control on steep slopes, compulsory dipping etc. Under existing legislation the Ministry of Agriculture has the power to take over and operate mismanaged or abandoned farms. This legislation has so far been implemented only in large farming areas, but its implementation should be extended throughout the country. The African society has the duty of imposing sanctions on those who refuse to contribute their fair share of work to the common endeavour [1]. However, the Government of Kenya is probably well advised to keep direct pressure on the producer restricted to a few items which can actually be controlled. Direct pressure as a general intsrument of agricultural policy is even more likely to fail in the future than it was under the soil conservation programme of the early fifties. Much, if not most, of Kenya's agriculture has already reached such a stage of technical progress that direct pressure is likely to be too crude [2].

4. A Possible Answer: The Combination of Incentives and Indirect Pressure

Considering the marginal economics of "close supervision" in the former Reserves, the limitations of the "voluntary approach" and of "di-

[1] African Socialism and its Application to Planning in Kenya. Nairobi 1965. p. 4.

[2] Since its failure in the early 1950's, direct pressure as an instrument of smallholder development is of bad repute in East Africa. It is not often enough realized that the failure was mainly due to the fact that the British Colonial Government was no longer in a position of absolute authority. There are a number of cases in agricultural history where direct pressure yielded quick and beneficial results when the peasantry was subjected to it. One case in point is Southern Korea during the Japanese occupation of Korea. Another case is the introduction of potatoes into Prussia by Frederick the Great. The success of direct pressure apparently depends on (1) the authority of the Government and (2) the nature of the innovation. The amount of acreage of a certain crop or the planting of a certain crop in rows can be controlled comparatively easily. But proper animal husbandry, fertilizing, proper weeding, pruning etc. — the items on which Kenya's agricultural development largely depends — are hardly suitable for control and thus for direct pressure.

rect pressure", it seems appropriate to Kenya's present situation to combine strong incentives with strong indirect pressure. In this connection two tools of agricultural development seem to be of paramount importance: (1) the judicious use of small loans to smallholders as a strong incentive and (2) the introduction of a substantial tax on the land as an instrument of indirect pressure.

a) Loans to Smallholders

The Government of Kenya is planning a substantial expansion of the loan programme. The distribution of public loans for agricultural purposes, which used to be in the hands of different bodies, has been concentrated in the Agricultural Finance Corporation (AFC). The only major field not covered by the A.F.C. are loans for the purchase of farms, which are given by the Land Bank. Smallholder loans are no longer taken care of by district personnel of the Ministry of Agriculture, but by loan officers of the A.F.C. Loans have proved to be a powerful incentive in support of the "voluntary approach". However, a policy of ample loans must be supplemented by the greatest strictness as to repayments. The attitude of most farmers to loans is that this avenue provides the panacea for their very problem. Rather frequently the efficiency in the application of money is low, sometimes because of bad luck but mainly because of poor husbandry. Unsuccessful smallholders usually feel that, as they used the

Table 30. *Statement of public loans to individual small scale African farmers since 1948*

Analysis as of 30th June 1964, in £'000

Source of loans	No of loans	Amount advanced	Amount repaid	Amount outstanding[1] more than one	less than year	Total[1] out-standing	Repaid as per[1] cent of total due
ALDEV	1336	77	25	12	3	15	62
ICA (American)	426	30	12	5	7	12	50
IBRD (World Bank)	3903	318	45	25	50	75	37
West-German	2185	179	3	.	9	9	25
Grand total	7850	604	85	42	69	111	48

[1] Principal and interest due as of June 30th, 1964. It has to be taken into account that the greater part of the amount due is outstanding less than one year and that some of it is paid with a delay of several months.

Source: Ministry of Economic Planning & Development, Nov. 1965, Mimeo.

loans without any tangible profits accruing to themselves, there is no rational reason why that loan should be repaid. Even the most sophisticated have only a limited knowledge of how a loan works, what interest means, and that repayments are inevitable and desirable from their own point of view. Few smallholders really expect to be forced to repay a loan by the Government. Public opinion is obviously not very strong in encouraging repayments of loans, and it can be seen that some of the areas which have relatively high cash incomes are the worst defaulters in respect of repayments. There is something like an optimum in the provision of loans to smallholders. If large amounts are distributed in a short time without the greatest strictness as regards repayments, it may well happen that the peasants concerned see a fair chance in trying to avoid repayments. After some time even those who are prepared to repay are discouraged to do so by the leniency towards their less responsible neighbours. In due course small loans to smallholders cease to be a useful instrument for rural progress. Experience indicates that smallholder development is not something which can be sped up by doling out ample funds and ample services in a short time, but something requiring a careful or even slow beginning and consistency over a long stretch of time.

The position as of June 30th, 1964, is depicted in Table 30. The bulk of the loans were only beginning to mature, but there were already indications that repayments would pose a great problem. In October 1964 the Government decided to restrict new issues in some districts in proportion to the percentage of arrears to gross issues. In districts with a percentage of 50 or more no further loans would be made for the time being; in districts with a percentage of 25—39, only 40 per cent of the budgeted allocation would be released; with a percentage of 10—24, 60 per cent of the allocation would be released; and with a percentage of less than 10 the full allocation would be released. As a result of these restrictions £ 70,000 has been frozen out of a total budget allocation of £ 276,500 for loans to individual farmers. The same problem arises in resettlement areas; the percentage of cumulative repayments in cash to cumulative debt service billings as of 31st May, 1965, was only 38. Commercial Banks which have been making loans (outstanding approx. £ 3,400) to smallholders of about £ 400,000 are faring somewhat better. But even they consider loans to smallholders an activity needing a certain amount of subsidy.

It is frequently argued that a good deal more education must be disseminated regarding loans. This is certainly true, but formal education is expensive, and the field staff cannot possibly advise every farmer in detail about the implications of a loan and how to use it. In any case, the effect of direct education or advice should not be overrated. No other way to educate the smallholder rapidly and effectively about loans is probably so quick and effective as the evidence that nobody has a chance to get away with not making repayments, that those who do not pay are brought to court, pay or lose some of their land. It is most encouraging that the Kenya Government is clamping down on loan defaulters [1]. There have been a few

[1] Rumour has it that there is local ostracism on the buyer, that bidding is rare and prices are low, i.e. that the neighbours buy the land at a very cheap price and give it back to the former owners.

exceptional cases where land has been taken away from loan defaulters. The same thing happened to hundreds of thousands of small farmers in Europe, the Americas and in Japan, i. e. everywhere where agricultural development took place within a framework of private ownership of land. Private ownership of land, as an institution which is expected to further the general welfare, can only fulfil its functions if the farmers are made to realize the full responsibilities which go along with it. Also, it must be realized that loans to small farmers rapidly lose their usefulness if loan repayments are poor. The decent farmer who fulfils his obligations will soon cease to do so if his less responsible neighbours get away with non-repayment.

One of the present difficulties with loan repayments is the fact that the co-operatives are not yet included in the procedure. Smallholder titles are only a last resort as loan security because of the difficulty and expense involved in cashing in on them. Experience in Africa generally indicates that it is expensive and difficult, if not impossible, to get the money back unless there is a one-channel marketing system. The handling of smallholder loans would be easier if the marketing co-operatives were included in the loan procedure in such a way that the instalments due are automatically deducted from the loaner's receipt with the co-operative. Obviously this will only be possible where co-operatives are firmly established and where there is a major cash crop which cannot find a black-market outlet (coffee, pyrethrum, tea).

b) A High Tax on the Land

Another instrument of even greater importance is the introduction of a substantial tax per acre. The Government paper on African Socialism already indicates steps in this direction by stating:

— Direct taxation on people with extremely low incomes will be progressively abolished and the revenue sacrificed replaced by property taxes.

It has been established by experience elsewhere that hardly any other measure is as efficient, as just in social terms, and as economic in promoting agricultural production as a high tax on the land. The argument that the "crushing" burden of school fees and taxes prevent smallholders from using their surpluses in ways which would develop their holdings, brought forward by almost every peasant in Kenya, is misleading. Smallholder development is connected with substantial costs for personnel, transport, Farmers' Training Centres, loans, soil-conservation etc. It can only be promoted over any length of time if it generates the revenues with which to finance public expenditure for these services. Thus, in a country like Kenya, substantial rural taxation by necessity goes along with a development policy which lays emphasis on rural development. Furthermore, it should be remembered that many improvements in smallholder farming are explicitly due to the fact that the farmers, being caught by the desire to earn money and to send their children to school, have no choice but to go into cash crop production. School fees and taxes are probably one of the most powerful driving forces of agricultural progress. There is, however, the justifiable argument that high school fees and the present poll tax are neither efficient nor "just" in social terms, and that a rather high land tax, substituting the poll tax and part of the school fees, might be more satisfactory. The main advantages of the land tax are the following:

— It would be a strong inducement to increase production per acre. The higher the value of output per acre the lower the percentage taken by the tax.
— A land tax is a light burden for the good farmer with high returns per acre, and a heavy burden for the poor farmer with low yields per acre.
— A land tax is a very heavy burden for the absentee landlord who puts his land into extensive grazing. Most of the undeveloped land in high potential areas would very soon be developed if there were a high tax on the land.
— The Government would obtain an important source of revenue with which to finance development.

The introduction of a high tax per acre certainly poses a number of difficult problems. It is probably the most unpopular tax of all because it hits hardest the most vociferous group in rural communities, — the somewhat larger farmers. A land tax can therefore be introduced only in a unified country where there is strong Government authority. Furthermore, there is the difficult question of how to vary the level of taxation so as to take into account differing soil fertility, climate, the transport situation etc. Perhaps a guiding line might be provided by the ecological zoning introduced by L. H. Brown. However, intricate as these problems might be, they have been solved elsewhere and there is no reason why a satisfactory answer cannot be found in Kenya.

In any case, the introduction of a high land tax must wait until all the high potential land of Kenya is consolidated and registered. The introduction of high taxation in areas already registered would put an automatic stop to the people co-operating in such matters as consolidation, enclosures and registration in areas where they are under way or have not begun yet.

c) The Importance of Land Registration

The effective use of both instruments, loans to smallholders and taxation of the land will be much easier where land consolidation has taken place. Land cannot be used as security and cannot be properly taxed unless it is registered and a title deed has been issued. Nobody is willing to spend money, time and effort on developing land without security as to occupation and tenure. It is important, therefore, that the highest priority be given to an acceleration of the process of land consolidation, registration and the issuing of title deeds to the land.

By the end of June, 1965, about 3 million acres in the former Reserves will have been consolidated or enclosed and demarcated, with registration completed on a little over half this amount. There are still 5.65 million acres of high potential land which have to be consolidated. Consolidation and demarcation is pushed ahead at an annual rate of roughly 350,000 acres, and at an average cost of approximately 20 shillings per acre of which 4 to 5 shillings have to be paid by the farmer (10 shillings in Central Province). A forthcoming survey will establish priorities as to the policies in the next years. The rate of registration, which is lagging behind, could perhaps be increased by putting more emphasis on areas already enclosed for some time. A limiting factor to the process is insufficient personnel, in particular survey staff. The demand on the part of the smallholders exceeds the ability to provide these services. By increasing the staff of the Land Consolidation Department and the finance allotted to it, the process of registration in the high potential areas might be accomplished — but this seems to be an optimistic estimate — within 9 years.

IV. Future Aspects of Agricultural Development in the Former Scheduled Areas

1. The Need for More Africanisation

Politics in Kenya have been revolving around the land and will continue to do so. The development of the high potential land in the former Reserves carries much more economic weight than any move in the Highlands, but the Africanisation of the former Scheduled areas is likely to remain in the foreground of agricultural policies. There was a remarkable degree of stability in Kenya in mid-1965, largely owing to the rapid progress in resettlement and the Africanisation of Government positions. It is our opinion that this stability will only last if Africanisation continues. It is certainly not in the interest of Kenya to abolish European or Asian farming altogether, but it is essential for the wellbeing of the minority groups themselves, for internal stability, and thus for African progress that the predominance of Europeans and Asians in certain sectors of the economy be replaced by a predominance of Africans.

The solution to this problem for the plantation industry and most of the ranching industry appears to be relatively clear: (1) the employment of African staff up to top-levels of management, (2) the purchase of shares and (3) the purchase of enterprises either by African individuals, co-operatives (for instance, coffee farms being bought by coffee co-operatives), companies, or the Agricultural Development Corporations. Politically advantageous as the Africanisation of the plantation and ranching industry might be, it is apparently not a question of great urgency since these industries use comparatively few acres which could be utilized by a land-hungry peasantry. On the other hand, their exports, employment and revenues are of decisive importance for the general economy. The process of Africanisation could probably be stretched over some period of time. This would avoid a disruption in management and a drop in production and would leave funds and personnel available for the Africanisation of other sectors of the economy.

Large-scale mixed farms owned by non-Africans present a much more intricate problem. They are politically more conspicuous than plantations and are more often adjacent to areas of increasing African land shortage. It was assumed in mid-1964 that there were 2.3 million acres of mixed farming land owned by Europeans and not yet earmarked for resettlement. Some European farmers have taken on Kenya Citizenship, showing their intention to stay. (Their land is estimated at 250,000 acres); some of them sold to Africans in 1964/65 (estimated at 200,000 acres); some land is farmed by commercial companies (estimated at 300,000 acres); and there is finally some land taken by larger residential plots (estimated at 150,000 acres). This leaves us with roughly 1.4 million acres which sooner or later will need a racial transfer in ownership. As a rough estimate it might be assumed that the purchase of this land will cost approximately £ 20 per acre, including loose assets, i.e. £ 28 million altogether.

Although the need for more Africanisation of mixed farming areas is accepted by all, opinions differ as to the scope and the speed at which the process should take place. European farmers campaign for a rapid and complete buy-out. They feel insecure. A farmer wants to work for the long-term future, which cannot be offered in Kenya. Election promises have been given by the African politicians to their land-hungry electorate. A slowing down in land purchases and land distribution might invite a land grab. Almost all of the European farmers look forward to vacating their holdings as soon as they can get an acceptable purchase price. Almost all African politicians are also in favour of a rapid Africanisation which, as far as the electorate is concerned, is the yardstick of their work.

On the other hand, it has to be considered that a complete buy-out of the European larger mixed farms within five years would certainly be difficult to contend with. Reduced speed in the take-over would allow for a more easy adjustment to the changes. The administrative burden would be lighter. Inevitably, the training of suitable personnel will take time. The British Government is particularly interested in a slow rate of take-over, first because it is likely to be asked for more aid if there are economic setbacks in Kenya, and second because more finance will be required for the purchase of the European farms and for resettlement and Land Bank loans to African large farmers or National Farms. Even more important is the likelihood that payments to Kenya farmers will create a precedent to payments in Rhodesia. Investment in European mixed farms in Rhodesia is approximately ten times greater than in Kenya, which would mean that Africanisation in Rhodesia would put several times the burden on the British Treasury than it has done in Kenya. Mrs. Barbara CASTLE, Britain's minister for Overseas Development, in 1965 stressed that according to the Kenya Government "any European farmer is welcome to stay as long as he identifies himself with the country". Mr. STAMP, leader of the Stamp Commission sent from London in early 1965, suggests that it would be a good thing for Kenya's economy if 30 per cent of the European farmers were to remain. He has been quoted as follows: "While I imagine that some further programme of transfer of European farms to Africans will be found desirable, it may well be that far more European farmers will have the possibility of leading useful and happy lives in Kenya than has been thought possible in times past." (Reporter, May 7th, 1965). Furthermore, there is the growing realization within the Kenya Government that the present practice of spending a large proportion of the Government's budget on resettlement, and of using Land Bank loans to support African buyers of large farms, is absorbing funds which could be used elsewhere. "What our country needs now is development, not merely land transfer." (T. MBOYA, Reporter, May 7th, 1965).

It is undoubtedly in the interests of Kenya that there should be more Africanisation in the former White Highlands, but that the speed of the process should keep pace with the organisation of an orderly transfer and that the form of the transfer should be as economic as possible and perhaps more economic than it has been so far. In this respect three points seem to be of paramount importance:

— Strong incentives must be given to the European farmers to farm properly. Apart from the maintenance of law and order, the most important single inducement for the farmer to run and develop his land according to the rules of good husbandry is the assurance that, sometime within the coming years, he will be offered a price which will take full account of the economic value of his holding. This is necessary to keep the former Scheduled areas in full production as long as there are European farmers in substantial numbers.
— There is the formidable task of stabilizing and fully developing agricultural production on the land which has already been Africanised. This is necessary in order to prevent the financial burden of resettlement and of African large farms from falling on the general economy.
— Appropriate institutions for the Africanisation of additional land have to be chosen. In particular the question arises as to whether there should be more resettlement, and if so, what kinds of schemes would be appropriate. Should there be more African large farms or National Farms?

The first point above depends primarily on the assurances provided by the British Government, which will in any case be responsible for buying out the European farmers. The second and third sets of problems, however, depend primarily on the action taken by the Kenya Government. It might be useful, therefore, to discuss certain aspects of possible proposals.

2. Stabilisation and Expansion of Resettlement Schemes

a) Stabilisation

In spite of all the difficulties of a transitional period, resettlement had a better start than was frequently expected and which leaves hope for the repayment of a significant proportion of the loans. However, the degree of success will largely depend on the policies of the next two or three years. A number of adjustments will have to be made in order to achieve the economic objectives of resettlement. There is the obvious need:

— to take unsuitable land out of settlement, to transfer the settlers concerned to other locations and to put the poor land into larger holdings[1];

— to weed out unsuitable settlers. There is a wide variation in the performance of settlers drawn from all sections of the community without any type of selection. As in all undertakings of this nature, some farmers are raising their standards and others are allowing them to fall. The latter, in particular those who do not pay their loan instalments, will have to be evicted if resettlement is to succeed;

— to replace the growing of maize on unsuitable soils and in cold climates by other crops. The growing of potatoes for subsistence and for cash, i.e. the establishment of a milk-pyrethrum-potato economy seems to be of particular importance in many areas;

— to provide the settlement staff with more authority in the supervision and control of land use and animal husbandry. Proper loan repayment can only be expected where there is proper husbandry; and this, in the present setting, is likely to be practised only if there is an element of pressure on the negligent farmers;

— to solve the problem of draught power. The co-operative use of tractors is likely to become rather expensive. In any case it will work only in locations with co-operative-minded settlers. The provision of loans for the purchase of tractors by successful settlers who will then do contract work for their neighbours may be the answer in some locations. Perhaps the Department of Settlement should reconsider the introduction of ox-drawn equipment. It will take much longer to stock the settlement farms fully with grade cattle than was expected, and furthermore, the growing of fodder crops can be made much easier by the use of ox-ploughs and carts, at least as long as not enough grade cows are available;

— to introduce innovations, new crops and new varieties etc. An important aspect is the introduction of fodder crops and the improvement of animal

[1] Land which was unsuitable for resettlement but which had to be bought by the Settlement Fund Trustees, and which is now being farmed by settlement personnel in larger units, should preferably not remain the responsibility of the Department for Settlement for any length of time. The difficult task of running farms for an interim period between purchase and resettlement has been handled remarkably well by the Department; but there is, however, the worldwide experience that state-owned commercial enterprises should not be run by Government departments but by statuary bodies because of the handicaps to the decision-making process which are, by nature, attached to Government activity. The farms in question, as well as the suitable personnel, could perhaps be taken over by the Agricultural Development Corporation.

husbandry. Extensive grazings are probably the most economic way of running a large farm on marginal land. Smallholders, however, are much more dependent on an intensive use of the land — which means that they will have to plant grasses and establish leys, in particular the clover-grass leys for high altitudes, in order to succeed.

Resettlement started with the idea that the functions of the Settlement Officer after some time would be taken over by the co-operatives and the relevant Government District Offices. However, considering the multiplicity of difficult tasks which the co-operatives are expected to perform and the intricacies of co-operative action, it is perhaps appropriate to keep the present institutional set-up for some length of time or even indefinitely — that is, to establish the Settlement Officer as the local representative of the benevolent landlord, the Settlement Fund Trustees. The provision of loans to the settlers can be considered a "power-key" which, if wisely used, can be of great benefit to the settlers. In our opinion then, the Settlement Officer should see that:

— the settlement rules are adhered to,
— the loans are properly repaid,
— the co-operative is functioning properly,
— the necessary soil erosion measures are carried out,
— proven innovations are speedily introduced into the scheme.

The power of the Settlement Officer should be comparable to that of the former landlord in Japan, or to that of the officers in many other schemes of "close supervision" in Africa. The landlord system has a bad reputation and understandably so. Most of the landlords — in particular in India and the Near East — (who have now been largely abolished by land reforms) showed the characteristics of a leisure class. However, the landlord system can be highly effective if the landlords are economically active, as they were in Japan. Peasant communities, in order to introduce innovations, need guidance and some degree of supervision, which can only come from somebody who has a "power key", i.e. who can revoke the tenant's contract or can cancel the loan. It is of course not advisable to establish private landlords in Kenya, but it is certainly worthwhile to consider the establishment of a landlord-settler system in specific schemes, with a statuary body as the benevolent landlord. There are numerous schemes in Africa where the Government or a statuary body took over the traditional functions of a landlord. A case in point is the organisation of the Mwea-Tebere Irrigation Scheme.

b) Expansion

In 1965 the speed of settlement slowed down because the Kenya Government wanted to focus its personnel and funds on the stabilization of the schemes already under way and to gain some time in order to assess their performance. The arguments against additional large-scale resettlement programmes in the next two or three years are powerful indeed.

— Most important is the shortage of personnel. The "Times" of January 12th, 1965 comments: "... in spite of the rapid recruitment of African staff, the appointment of highly competent Africans to a number of senior posts, and intensified training of scores of younger men, the present scheme must depend for a few years more on the dedicated efforts of a group of expatriate officers, many of them former farmers... An addition to present settlement would place an even heavier load on this central core of experienced officials since it would double and treble the need for administrative skills, particularly African skills, at a time when the existing scheme already strains effective administration to the limit... The plain fact is that some 70 per cent of these key men must leave in 1965. Their contracts are running out. The Kenya Government has neither the resources nor the political elbowroom to offer them the only guarantee that can keep them in Africa — the guarantee of a life's career in the agricultural field..."
— Additional settlement will encroach more and more on highly developed large farms, where it will be more difficult to exceed the present marketed sales per acre than in the current schemes of which large parts have been underdeveloped.
— Skilled personnel and funds are needed for other development schemes, the economic benefits of which are greater than those of resettlement.

If political criteria leave no other choice but to continue with further large-scale resettlement, a two-way approach might be considered appropriate. So far experience has shown that ample loans and efficient personnel do not suffice to turn marginal land into good settlement areas. If the resettlement of marginal land cannot be avoided, it is perhaps more economic to hand it over to settlers at a rather low purchase price and to let them take care of it themselves without additional specific development loans.

High potential areas, on the other hand, deserve planned resettlement. There will always be a divergence of opinions between the agriculturalists and economists on the one hand, who want to have settlement for economic growth and who consequently favour the promotion of Low Density schemes, and administrators and politicians on the other, who want to settle as many of the landless as possible and who consequently wish to promote High Density schemes. There seems to be a consensus of opinion in Kenya that Low Density schemes offer better chances for the repayment of loans, and should be preferred if priority is to be given to economic development. If, however, short-term political effects are considered to be more important, High Density schemes will have to be chosen. Additional resettlement would probably cost about £ 30 per acre; thus a programme of another 600,000 acres would amount to £ 18 million. The question as to whether or not it will be worthwhile to have another programme depends on the loan repayment and production performance of the present schemes.

3. Stabilisation and Expansion of the Large Farm Economy

a) The Need for Large farms

Although resettlement was necessary and is likely to continue, it is not expected to lead to the complete partition and distribution of the mixed farming areas among smallholders. There are strong arguments against the preponderance of large farms in the early stages of economic development and in particular in countries where rural underemployment is rampant; but there is also much to be said against concentrating exclusively on small farms. The best solution in most circumstances would be the combination of farms of different sizes within the country. Lord Delamere's original idea of the "White Belt" in the middle of Kenya has been passed by history, but the idea of having a belt of "modern large farms" surrounded by smallholdings has much to recommend it in the present circumstances.

Table 31. *The impact of change in the mixed farming economy in the former Scheduled areas*

	Wheat '000 Acres Large Scale	Settlement	Total	Maize[1] '000 Acres Large Scale	Settlement	Total
1959	254	—	254	135	—	135
1960	248	—	248	142	—	142
1961	226	—	226	158	—	158
1962	244	—	244	159	—	159
1963	278	5	283	112	8	120
1964	282	8	290	75	26	101

	Dairy Cattle[2,3] '000 head Large Scale	Settlement	Total	Beef Cattle '000 head Large Scale	Settlement	Total	Sheep '000 head Large Scale	Settlement	Total
1959	428	—	428	532	—	532	549	—	549
1960	413	—	413	567	—	567	583	—	583
1961	393	—	393	549	—	549	580	—	580
1962	356	—	356	530	—	530	511	—	511
1963	332	10	342	481	12	493	502	12	514
1964	285	44	329	452	12	464	409	31	440

[1] The decrease in the planting of maize was largely due to the drop in prices and the increase in theft.

[2] The reduction in dairy cattle is partly due to the weeding out of low yielding cows.

[3] Some of the grade heifers and cows have been sold into the Reverve, where the number of grade cattle is estimated at 50—60,000.

Source: Ministry of Agriculture, Nairobi, Mimeo, 1965.

Kenya's belt of large mixed farms has been reduced by resettlement and has suffered in quality in the wake of political change. There is a marked decline in deliveries of pyrethrum. The acreage of maize has decreased significantly and this is obviously related to resettlement. In addition there has been a steady decline in the number of dairy cattle, wool sheep and breeding pigs. (See Table 31.) But there still exists a sizable block of large farms which should be kept and fully developed for the following reasons:

— Substantial acreages in the Highlands are not suited to smallholdings because of heavy soils, low potential or an adverse climate.
— Certain agricultural activities which are highly important for the internal and export market do not fit easily into smallholdings; the smallholdings will probably produce increasing sales of pyrethrum, potatoes and milk; but the maintenance and expansion of market sales of wheat, barley, oats, maize, wool, mutton, beef etc. will depend on the existence of a belt of large mixed farms.
— The bottlenecks of smallholder development can be overcome much more easily with a division of labour between small and large farms. Smallholdings are badly suited to the production of seed grains, healthy potato-seed, in-calf dairy heifers, improved in-lamb ewes, disease-free in-pig sows etc, while these activities fit well into large farms.
— Sustained development in agriculture is highly dependent on a large number of innovators. Experience in Kenya, as elsewehre, clearly shows that smallholders and Government officials are comparatively poor innovators. Research personnel always have the tendency to go into interesting problems which are of little practical relevance. There is no better innovator than a commercially minded large farmer, be he the owner, tenant or a manager of some independent standing. A belt of large farms is necessary in order that a flow of innovations may be forthcoming.
— Applied agricultural research in Kenya, being pushed by the large farming community, has made a "jump ahead" recently. e.g. rust-resistant wheat varieties, hybrid maize, clover-grass mixture, proper sheep husbandry etc. The introduction of these innovations into farming in the former Reserves and in resettlement schemes will be slow. On the large farms, however, they are being rapidly applied on large acreages. Considering both these innovations and the tremendous potential for more intensive land use on most large farms, it is quite conceivable that, provided proper management is available and law and order are maintained, the mixed large farms, although having lost more than a third in acreage, might reach and surpass the output of 1962.

The above functions of large mixed farms would, in normal circumstances, best be performed by their present owners who have the capital, the credit and an intimate knowledge of their holdings. As things are, however, they will certainly not develop their holdings as before. In any case, more Africanisation of the land is urgently required. Several basic types might be visualized:

— African large farms of different sizes run by individuals, partners or companies;
— co-operative or collective farms;
— National (State)-farms.

By mid-1965, the Kenya Government had not committed itself to any particular form of ownership and management of large farms, but a great number of different approaches were discussed and actually tried.

b) A New Approach to African Large Farms

Additional financial and technical support for African buyers, be they individuals, partners or members of a company, is one way to preserve the belt of large mixed farms under the changed circumstances. This approach to Africanisation is likely to be cheaper than resettlement or National Farms.

Table 32 gives a rough estimate of the public costs of different approaches to the Africanisation of mixed farm production. African large farms have the advantage that part of the necessary capital is provided by the new owners' private sources. The percentage of the required capital which is mobilized by the settlers of Low Density Schemes is significantly smaller, and with National Farms it is nil, i.e. more public funds

Table 32. *A rough estimate of the economics of different approaches to the Africanisation of mixed farming areas*

	in £'s per 1,000 acres Low Density Resettlement	African large Farms	National Farms
Land purchase[1]	11,000	11,000	11,000
Land development[2] and working capital	11,000	6,600	11,000
Planning, overheads, administration, fees etc.	8,800	398	2,000
Total	30,800	17,998	24,000
Farmer's own capital	2,800	4,498	—
Public loans required	28,000	13,500	24,000
Assumed annual marketed[3] production before change in ownership	4,700	4,700	4,700
Assumed annual marketed production four years after the change in ownership[4]	7,050	3,700	4,700
£ public loans required for each £ marketed produce	4.0	2.9	5.1

[1] The calculation is based on the assumption that the quality of the land is the same in all three cases. This assumption is not likely to apply. National Farms will probably be established on highly developed land where land values are higher than for land bought by African large farmers or by the Settlement Fund Trustees.

[2] Capital requirements will be high for settlement because of additional feeder roads, water points, soil erosion works etc. Capital requirements on National Farms will be roughly as high as or even higher than those on properly run European farms, while African large farmers will try to get along with little capital.

[3] Average figure for European large farms.

[4] The calculation is based on the assumption that the marketed produce of Low Density Resettlement will be 50 per cent above, that of African large farms will be 20 per cent below, and that of National Farms on the same level as the former performance.

are required to Africanise production even if we assume — as is done in Table 31 — that marketed production on African large farms will be 20 per cent below the former level. On the other hand, the general economy has to forego the urgently needed production because the buyers lack sufficient capital and experience to develop the production potential of their land. Some changes in the policy towards African large farms might improve the situation.

Priority must be given to the stabilization of production on farms which have been bought in the last three years. A salvage operation, i.e. the employment of additional extension officers for African large farms, began in mid-1965, partly with personnel supplied from Germany. Supervision of and advice to new farmers will certainly be of value. It should not be overlooked, however, that without the "sword of Damocles" of foreclosure hanging over the farmer, the advice of any extension officer will meet with some reluctance. Foreclosures of those who do not follow the advice and do not pay their loan instalments will have a much better educational effect than credit coupled with friendly persuasion. The Land Bank is embarking on a programme of employing full time agents — in addition to the personnel supplied by the Ministry of Agriculture — who will undertake extensive inspection work on large farms. All these efforts which are rather expensive, will need a certain number of foreclosures in order to show results, and hence they will need the full backing of the politicians.

Large farmers have obtained an enhanced position in the rural economy with the help from public funds. Such support can only be justified if they make proper use of their opportunites. It is certainly useful to the economy to have and to support African entrepreneurs but there is no point in supporting an African leisure class.

Some of the problems faced by African large farmers could perhaps be avoided in the future by a change in policy. A crucial shortcoming is the lack of working capital, and this might be eased by a two-way approach:

— The Land Bank could insist (1) that the purchaser has more than 50 per cent of the purchase price of land and permanent improvements available in ready cash [1], (2) that the availability of native cattle is no longer recognized as an asset, and (3) that the buyer deposits his working capital with the Land Bank, which will make the money available only against certified invoices covering productive work and purchases on the farm. It would, furthermore, still be necessary for the Bank to act quickly and foreclose if a mass movement of relatives and friends onto the farm showed that the loan applicant had obtained his capital in return for undefined and illegal squatter rights on the property.

[1] Out of a total of approximately £ 18 per acre required to buy and run a larger farm, the Land Bank would provide £ 7, the AFC £ 3 and the buyer would have to find £ 8. Up to now the buyer has only had to find £ 4.

— For many years the European Agricultural Settlement Board pursued a policy of buying large farms and breaking them up into smaller units. This policy resulted in greatly increased production and it can certainly be continued. Clearly, there is considerable scope for a reduction in the size of the present mixed farm, which averages 1,350 acres, to units of roughly 400 acres total acreage and 200 acres arable land, which could use tractors. The basic pattern of mechanized farming would not be destroyed, a greater number of Africans could become larger farmers, the capital of the individual would go farther and there would be less danger of relatives and friends encroaching on the land. However, the sub-division into tractor-sized units poses some problems: additional water supplies, buildings and roads would have to be constructed; and it would be more difficult for the Bank to supervise the farmers. But the advantages of such a move seem to outweigh the disadvantages.

Corresponding changes in the loan policy of the Land Bank would probably result in (1) more Africans being settled but (2) less land being Africanized. This, in our opinion, need not be considered a serious disadvantage. In any case, only a part of the mixed farming areas could possibly be bought by African individuals, partners and companies. Other solutions to the problem must be found for the remaining land.

c) Co-operative Farms [1]

Another approach to the Africanization of large mixed farms in the Highlands is the establishment of co-operative farms, whereby the members — usually farm workers — initially forego the profits in order to buy themselves into the ownership of the farm.

There is much advocacy for co-operative farming which in theory seems to have the ideal combination of economic and social advantages [2]. But experience and evidence does not confirm the theory. While co-operative marketing has a long record of success, there are few examples, if any, of successful co-operative land use in East Africa. Kenyans, and in particular the Kikuyu, are notorious individualists who shy away from the discipline which is a vitally important aspect of co-operative land use. The task of the secretary of the co-operative more often than not is frustrated by committees who seem reluctant to accept the principles of co-operation. Successful large-scale farming, in particular mixed farming, obviously implies (1) a great number of daily on-the-spot decisions, which (2) can reasonably only be made by just one person who must have the necessary experience in deciding what should be done, who (3) carries the full responsibility and (4) whose economic success (profit, salary, promotion) must be clearly

---

[1] Types, problems and experience thus far with group farms and co-operative farms are dealt with by N. Neviger, whose report is in print.

[2] See: Kuvshinow, I. S., in: Proceedings of the International Conference of Agricultural Economists, London 1963, p. 307. — Rosenkranz, O.: Zur Problematik der Betriebsgröße in der Landwirtschaft. Symposium der landw. Fakultät, Leipzig 1959, p. 14 ff.

related to the profit of the enterprise. The conditions are difficult to fulfil as far as co-operative land use is concerned. Reviewing the numerous efforts in co-operative land use which we have seen in East Africa since 1961, we are inclined to believe that it is likely to succeed only if the members of the co-operative have no influence whatsoever in the running of the enterprise. It is certainly appropriate in Kenya's present situation that some plans for pilot co-operative land use are started; poor performance in the past does not necessarily imply a failure in the future. It should not be overlooked, however, that there is a likelihood of the land of the co-operative farm, after an interval, being subdivided into subsistence plots, and that as a consequence loans will not be repaid and productivity will not be maintained. It may well happen that the establishment of co-operative farms will sooner or later prove to be a disguised and expensive form of unplanned High Density Settlement. If this is the likely outcome, it seems that the land should have been given to High Density Settlement in the first place.

d) National Farms

When we consider the necessity for transferring more of the Highlands of Kenya to African ownership, and the costs and difficulties of resettlement, the lack of experience of African large farmers, and the notorious pit-falls of co-operative land use, it seems that state ownership of the land, and its management by an Agricultural Development Corporation — a statuary body — might be a possible solution to some part of the problem. In mid-1965 an Agricultural Development Corporation came into existence in Kenya. It has the power "to raise money, to lend or to advance money to agricultural development projects, to acquire interests in agricultural undertakings and generally to promote and manage schemes". The absence of such a body in the past has been one of the major impediments of development in the large farming areas. It is now designed to take over land from outgoing European farmers. Approximately 100,000 acres of abandoned or mismanaged farms were in the process of being taken over by the Ministry of Agriculture in mid-1965. It was assumed then that there might well be a further 100,000 acres in a condition which would warrant similar action. This would mean state ownership of large stretches of land (although only a small fraction of the high potential land would be taken up) and the operating of part of the land as "National farms" by employees of a state-owned body.

National Farms do have certain advantages: the land would be Africanised; the belt of large farms would remain intact; National Farms would probably be a somewhat cheaper way of Africanising the land than resettlement; and the efficiency of land use is likely to be greater than on African large farms. They are, however, no ideal solution. Their purchase and development would require much capital. Scarce personnel would be absorbed for the substitution of outgoing European farmers. People in the former Reserves would be likely to show the same opposition to large-scale National Farms as to large-scale European farms, because in both cases land is occupied which they want to buy or settle on. The varied character of the mixed farms with their different types of problems would make it difficult to avoid losses. State-operated mixed farms have a poor record in many countries. Historical evidence shows that hardly any Government can resist the temptation to interfere in the running of farms which are owned by the state or one of its statuary bodies. Any reader of the relevant discussions in the House of Representatives will realize that the chances of political non-interference in the running of large scale state farms are very slim indeed. The managers usually do not have enough freedom to make decisions and are given too few incentives. More often than not the inflexibility of any administration and the red tape which usually results from state ownership leads to laxity, belated decisions, and an accumulation of losses which have to be shouldered by the Treasury.

However, the pitfalls of state-owned farms can be avoided. They are not inevitable but are usually due to unrealistic Government measures. One of the prerequisites of economic success is that the Agricultural Development Corporation should be a money-making concern run on strictly commercial lines. The actual farming operations should certainly not rest with any Government department. Given a reasonable level of management, sufficient working capital and a substantial element of self-interest in the payment of managers, production could be expanded and sufficient profits could be made, with which to pay back the purchase price. There is no reason why incentives to maximum efforts should not be built into the salary structure. Qualified managers could be employed regardless of where they came from. It would certainly be more acceptable to have Europeans as managers of National Farms than as farm owners. Even if National Farms were not as efficient as private farms under normal circumstances, it must be taken into account that under prevailing conditions private farms cannot be expected to show their best performance. If the Agricultural Development Corporation builds up the reputation of a successful commercial undertaking — as has been the case with the Kenya Tea Development Authority — it might become a good institutional framework for attracting capital aid.

If experience shows that National Farms run into losses, the Agricultural Development Corporation should have freedom to lease farms to large-scale tenants, be they Africans, Europeans or Asians, individuals, groups or co-operatives. The land would remain vested with the Kenya Government. The European or Asian entrepreneur would certainly be more acceptable to African Socialism than the owner-farmer; and the capital of the African entrepreneur would go much further if he were a tenant than if he were an owner-farmer. Of particular interest might be the leasing of land to commercial firms with a special interest in certain commodities: wool, seed potatoes, milk etc. If the National Farms proved too costly an undertaking, it might be acceptable to lease out most of the land. The land rent paid by the large-scale tenants might, in time, become an important source of revenue to the Treasury [1].

V. New Aspects of Land Development

Beside the smallholder and the large farming areas, Kenya has vast stretches of land which are used, in the main, only by herdsmen for extensive grazing. Most of it is suitable for ranching, and could yield two or three times the present gross return. Some land, particularly in the Masai Reserve, could be opened up for dry-land farming. Moreover, there is substantial scope for irrigation development. The possibilities of land development should certainly not be overrated nor the difficulties underrated. The modernization of animal husbandry in the semi-dry areas seems to be one of the most intricate social problems in Africa. The scope for additional dry-land farming is rather limited and depends on the co-operation of the relevant tribes. Irrigation development is expensive and takes much time to mature. In any case, the increase in output per acre in the present smallholder areas and the former Scheduled areas is far more important than additional land development.

1. Grazing Development

The deterioration and complete breakdown of most grazing schemes established in British times has had its impact on the pastoral tribes and on the Government. The herdsmen increasingly realize that nomadism and progress exclude each other, that children can only be sent to school and other social services be enjoyed if the population is stationary. These considerations, together with the general social change brought about by Independence, are apparently opening minds to reform. All of a sudden, dipping — which used to meet with fierce opposition during the colonial rule — has become popular. It would appear that even the Masai are changing their old ways.

[1] In Nineteenth Century Prussia, for instance, the state owned approximately 10 per cent of the best farming land and rented it out to large-scale tenants. The advantages were manifold: the state received important revenue from the tenants; it largely controlled the market for land; agricultural entrepreneurs with little capital had a chance to start as large farmers. In fact, agricultural progress has largely been carried by these large-scale tenants.

Another amazing change is the sudden movement for enclosures among several pastoral tribes. It was the policy of the Kenya Government that small individual ranches should not be allowed except on the very best range land. However, the tribesmen concerned did not ask for Government permission, but started to divide the land among themselves. The movement for enclosures of semi-arid land is particularly strong in the Masai country. The Masai Councils actually allocated approximately 100 ranches of between 500 and 2,000 acres to individuals without bothering much about their uncertain legal standing. The allocation of individual ranches by the relevant Masai Councils has been stopped temporarily in 1965. However, Masai politicians are urging that these farms be demarcated and be legally recognized by the Land Department. Requests have been received by the Minister for Lands from the Masai elders for work to be started on the survey of the whole Masai country, with a view to the issue of individual titles. The trend has been set by the more advanced and educated Masai, who believe that enclosures are the prerequisite of progress. The movement for enclosures gained momentum in 1964, and vast areas of hitherto communal grazing lands were allocated by the County Councils. It was believed, in mid-1965, that one of the two Masai districts, Kajiado District, had already been completely subdivided by the tribe with the members of every "manyatta" (group of families) knowing the exact borders of their land.

On the other hand, there is a change in Government policy. A special Range Management Section has been established in the Ministry of Agriculture. The emphasis is no longer on administrative action in order to prevent erosion as in the former grazing schemes but on (1) economically promising locations, (2) the allocation of land rights to allow for the development of permanent water supplies, which should stabilize herd movements and may later lead to fencing and (3) the money-making aspects of livestock management. The approval of funds and staff for each scheme is made dependent on the acceptance by the people of the following conditions:

- an indication of a real desire for development and the willingness to co-operate;
- an agreement to some form of legal land title or boundary agreement, whether the title be held by individuals, companies, co-operatives, associations or family groups;
- an agreement to comply with the rules attached to a specific range development scheme;
- an agreement to pay the stipulated rates and cesses.

The support or pressure for grazing schemes as provided in the late fifties has been replaced by a more diversified approach:

Government ranches. There are still some 5.5 million acres of unoccupied land, largely on the coast, not used by pastoral people because of tribal boundaries, lack of water, tsetse infestation etc. It is assumed that 3.8 million acres of this land are suitable for ranching, and that the Agricultural Development Corporation, possibly in partnership with private firms or entrepreneurs, could develop and use these areas.

Individual African ranches. The principle of individual ranches has been accepted where it is the urgent demand of the people concerned and where the unit is sufficiently large and provided with water to be developed economically. This approach does not receive particular encouragement. The allotment to some individuals of sufficiently large ranches would mean that many others have no place to live. Indeed, the division of the whole area would give hardly more than 300 to 400 acres of suitable range land per family, which is far too small a unit in most locations. L. BROWN writes on this subject: "It is generally accepted that to be economically viable a unit of range land must be large or very large — the bigger the better as a rule ... Given that the methods of management in each case (500 acres and 10,000 acres) involve perimeter fencing, five paddocks to provide a minimum of control, water supplies in each paddock, a dip and a minimum amount of necessary buildings, the costs per acre work out at about 10 shillings on a 10,000 acre holding and 40—45 shillings or even more on the small ranches of 500 acres. In both cases the ultimate gross return per acre which is to be expected is around 9—10 shillings under optimum management."

Group ranches. In the majority of the range areas economical and ecological considerations will force the people to come together on the basis of clans, families or other acceptable groupings and to establish ranching units which might comprise two to twenty families and an acreage of between 1,000 and 30,000 acres. So far the individual generally exercises grazing rights over the entire area recognized as the grazing grounds of his clan. He thus has, along with other members of the clan or tribe, freedom to move with his herds over a considerable area, but has no specific individual rights to a particular portion of it. The new move would mean the division of the area into self-contained units as far as water is concerned. Only people belonging to a certain group of families would then have the right to graze there. The ownership of the land could go to the group or clan, provided that this could be legalized. Another perhaps more realistic approach would be the establishment of group ranches on the land of several individuals. However, in each case the individuals would retain the ownership of their cattle. Some of the Masai, for instance, are reported to have decided to divide their land into units of over 20,000 acres to be owned by family groupings whose interests are homogeneous and who have traditional respect for the elders in the group. It is intended that the land of each ranch will be registered in the participants' names and that a management committee will be formed whose function will be to run and develop the ranch, taking up loans when necessary.

Co-operative ranches. As a next step in institutional development the Government intends to urge — as far as is feasible — that individuals and family groups lump their grazing and their cattle together and establish ranching co-operatives. These are expected to be run by a manager and to raise their own funds. The individuals concerned may receive shares according to their contribution. They might be employed by the ranch management as is convenient. Actually the name "co-operative" is misleading. The institutional set-up is much closer to that of a company than of a co-operative.

Location schemes. This is a new name for the old "grazing schemes" which are to be revived wherever the other propositions are unlikely to work. It is considered that these schemes can be restarted successfully if further improvements are made in water distribution and if the politicians render support.

The new diversified approach is probably the most comprehensive, most pliable and thus most interesting approach to the intricate system of land use in semi-dry areas. Government ranches are likely to offer good prospects as long as management is sound and the politicians do not interfere. The next years will show whether the relatively small individual ranches will be viable and will bring improvement in husbandry methods. The efficient working of group ranches — which are expected to cover more than 7 million acres — is certainly open to some doubt. The ranching co-operatives or companies seem to be somewhat more promising, because the responsibility does not lie with a committee but with a manager whose powers could be clearly defined.

In any case, — and here again we turn to L. BROWN (p. 29) — whatever the objectives of aid, two basic requirements must be fulfilled if money is to be invested in pastoral areas:

- limiting the number of stock which attempt to graze any area to the number which that area can carry without deterioration over a long period of time;
- controlling the movement of stock within any given area through rotational grazing systems or other such measures designed to maintain and/or improve the quality of the pasture land.

Without these two basic practices no other range improvement measure — whether additional water supplies, tsetse clearing, reseeding, or bush control — is of the slightest value. This is also true of veterinary services, which — except with devastating diseases like rinder-pest — can only be of economic value within the framework of a rational system of land use.

Finally it must be pointed out that the agricultural problems of most semi-arid areas can be solved only by the removal of a great number of people from the pastoral way of life. There are too many herdsmen who try to live on their stock, and who therefore attempt to retain, against advice, as much stock as they can for their own survival, thus continuing the destruction of their habitat. The relative degree of overpopulation is usually higher than in densely populated high potential areas. But experience everywhere in Africa shows that it is extremely expensive to settle pastoral people as irrigation farmers. There is, to our knowledge, no other answer but sustained efforts to provide urban employment.

2. New Land for Arable Farming

The possibilities of opening up additional rain-fed land for arable farming are rather limited in Kenya. This is playing a minor role in the development programme. The poor results of the former Aldev settlement schemes on marginal land have certainly contributed to the present cautious approach.

The clearing of land in expansion areas, carried out by peasants moving from densely populated hills in drier areas, is evidently going on. The development of cotton as a cash crop, the growing interest in chillies, castor and sisal, and the recent availability of early maturing drought-resistant maize varieties made it worthwhile to clear land hitherto considered unsuitable for arable farming. More planned settlements are likely to occur at the coast and on some tsetse-infested pockets round Lake Victoria. It is thought now that settlements should not be considered as an instrument of social relief but as nuclei for developing larger areas, that considerable initial development will have to be carried out, that close supervision will be introduced, and that all costs will be recovered from a development charge. However, in both areas the economics of planned settlement are likely to become difficult. The opening up of the land or the clearing of tsetse bush will become expensive. "You give the settler a lot of bush and forest among elephants, and he has no machinery to help himself with ... He manages to clear half an acre for the elephants to eat and destroy ... It is quite useless to dish out settlement plots on that basis. The Minister must give machinery, initial assistance to the settler to go ahead with", is argued by Mr. Ngala in the House of Representatives (13th July, 1965). On the other hand — considering the areas in question — it is most unlikely that the development costs can be recovered from the settlers by loan repayments or cesses on the produce. There is always a danger in East Africa that settlement on newly opened land takes place prematurely, i. e. before a proper type of farming has been developed and some experience as to the possible budget of settlement farms has been gained. Wherever feasible, a nucleus large farm in the centre of the area in question might be a cheaper way of developing land — although the large farm is likely to run into losses — than prematurely planned settlement with a lot of hidden subsidies which obscure the real costs of the scheme.

Land development in the Masai Districts is a different proposition. There are 200,000 acres of high-fertility, high-rainfall land well suited for wheat and woolled sheep at Mau Narok, presently used for extensive grazing, and there is probably more arable land in other pockets of the two Masai Districts. The development of these areas, which preferably should not be settled but used as larger, mechanized farms — a task which could be tackled by the Agricultural Development Corporation — depends mainly on the willingness of the Masai tribe to sacrifice grazing grounds, to grant distinct land rights to the agency which will be carrying out the development work, and to change to arable farming or to accept compensation payments. The agricultural and economic conditions of land development at Mau Narok seem to be favourable. The area in question is similar to some of the best wheat country in the former Scheduled areas. The problem is one of finding an institutional framework which will be accepted by the tribe and which safeguards sound management. According to the experience in East Africa so far, sound management will not be possible as long as tribal or local committees have a decisive influence. A feasible solution might perhaps be a company, with the tribe which is contributing the land, the Agricultural Development Corporation and an aid-giving agency as partners, whereby the majority vote is not with the tribe.

3. Irrigation Development

While the scope for increasing the cultivated acreage of Kenya by the extension of dry-land farming is limited, important possibilities are open for irrigation development. L. H. Brown (p. 43) estimates that

Table 33. *Plans for irrigation development*

Location	Irrigation [1] Potential	Actually [2] Irrigated in 1965	Irrigation [3] Planning, 1964—1970	Estimated [3] Expenditure 1964—1970
	acres	acres	acres	£
Mwea-Tebere	12,400	7,400 [4]	5,900 [5]	432,000
Tana River	75,000 [6]	1,200	60,000 [6]	3,500,000
Yala Swamps	20,000	—	500 [7]	324,000 [7] 1,500,000 [8]
Kano Plains	30,000	—	500 [7]	50,000 [7] 500,000 [8]
Taveta	7,600	—	1,100	70,000
Yatta	1,000	—	1,000	132,000
Perkerra	6,500 [9]	1,700	?	300,000 [10]
Kerio-Turkwell	12,000 [11]	—	2,000	300,000 [12]
Total	164,500	10,300	.	.

[1] Only irrigated areas under consideration.

[2] Under organised schemes only. There is also local irrigation, particularly in the Kano Plains.

[3] According to 1964—1970 Development Plan.

[4] Including 2000 acres, the development of which began in 1963/64. Britain's Freedom from Hunger Committee is to furnish aid to extend this irrigation scheme by another 2,000 acres, and by 1970 the Kenya Government hopes to bring the figure to 12,000 acres.

[5] Including 900 acres of red soil.

[6] Preliminary; the area is under investigation.

[7] Pilot scheme only.

[8] Planned expenditure for the planning period.

[9] Provided an expensive storage tank is built.

[10] Token provisions in the plan.

[11] Preliminary, depending on further investigation.

[12] Token provision. Other schemes have priority.

Sources: Kenya: Development Plan 1964—1970, p. 60; Brown, L. H.: A National Cash Crops Policy for Kenya. Nairobi; 1963. p. 43; Brown, L. H.: The Development of the Semi-arid Areas of Kenya, p. 36, 41.

583,000 acres could be irrigated in Kenya, while other offhand estimates run as high as 1 million acres. Development thus far has only touched the fringe of the potential. The total area of land irrigated under organized schemes in mid-1965 was 10,300 acres (See Table 32). In addition there are some areas with overhead irrigation in the former Scheduled areas, particularly for coffee, and roughly 10,000 acres of African irrigation by local methods.

Kenya has reached a stage in development where large irrigation schemes are urgently needed:

- There is a growing land shortage and little unused land that can profitably be farmed intensively without irrigation.
- There is a growing internal market for the produce of irrigation farming.
- There are more and more farmers' sons who cannot be absorbed elsewhere, who have a background of arable farming and realize the necessity of putting great efforts into their performance as irrigation settlers.
- Irrigation development and farming has a rather high employment content.
- Kenya is highly vulnerable to the vagrancies of an unreliable rainfall. Food production in particular is strongly influenced by the variation in weather. Irrigation development would bring some stability into the farming industry and could substantially lower the costs of famine relief in dry years.
- The task of irrigation development could largely be transferred to private enterprises, and the additional burden on Government departments could thus be kept low. The staff demands of irrigation farming are low compared with extension: in irrigation farming the subordinate staff receive clearcut orders which have to be executed and it is far easier to obtain suitable staff for such a purpose than for extension.

The need for irrigation development is realized in the Development Plan for 1964—70, which envisages the irrigation of an additional 71,000 acres, particularly along the lower Tana river.

Future irrigation development in Kenya is likely to be more expensive than in many other parts of the world. There are undoubtedly numerous possibilities for small local schemes which do not offer sufficient scope for the organisation of a proper scheme and for the employment of a highly qualified manager. These possibilities could be developed cheaply if the peasants concerned would be prepared (1) to do unpaid communal work in order to establish the necessary works, (2) to submit themselves to strict water control, (3) to farm the area properly and (4) to pay a water rate sufficient to repay the initial investment. All this has been customary in East Asia but it is certainly not likely — at least within the present setting — to occur in Kenya.

The establishment of large irrigation schemes is necessarily costly. According to L. H. Brown, overall capital expenditure in Kenya is generally in the order of £ 100 per acre without any storage. The costs of storage dams are likely to amount to another £ 100 per acre. Combined storage and development costs will thus add up to £ 200 per acre. These estimates seem to be on the optimistic side, considering that most of the irrigation potential is situated at the Tana River. Land development around the Tana River is likely to involve but moderate costs. Water procurement, however, will become very costly. The Royal Commission Report of 1955 already pointed out:

— From the technical point of view, therefore, it appears that the prospects are limited by the enormous costs (per acre) involved;
— the Tana River has certain other characteristics which must be taken into consideration when the possibility of the economic use of irrigation is under review. The riverine lands are remote from the centres of population;
— there is the virtual absence of communications;
— the site is two hundred miles from the nearest railhead;
— Hill dwellers from the congested land units in the Highlands might well be reluctant to emigrate to the hot and extremely unhealthy Tana River valley (p. 272).

The conditions for production at the Tana River make it difficult or even impossible to find a system of land use the returns of which would be sufficient to carry the burden of amortisation and interest of an investment which might add up to £ 300—400 per acre.

F. Conclusions

Kenya is at a particularly vulnerable stage of economic development:

— The agricultural economy relies heavily on the activity of minority groups which will have to give way to Africanisation. In Tanzania and Uganda the dependence on European or Asian enterprise is far less pronounced.

— Rural areas are full and the population is increasing more rapidly than elsewhere in East Africa. There is less free arable land which can act as a sponge to absorb surplus population than in Tanzania or Uganda.

— A significant percentage of the population, and one much higher than in Uganda or Tanzania, has been brought into the exchange economy, either as labourers or as cash crop farmers. Thus the traditional contentment with things as they were has largely disappeared.

— The only major industry to rely on is agriculture.

— The returns of agriculture are insecure because of the dependence on an unreliable rainfall.

All this indicates that Kenya needs aid more urgently than do most other countries in Africa. Moreover, it offers a number of advantages:

— There is an ample, comparatively industrious labour force adapted to wage labour.

— The basic requirements of an infrastructure are available. The services needed for agricultural development have largely been established and kept in working condition.

— The planning, particularly on the technical side of agriculture, is comparatively precise and realistic.

However, there is reason for concern — and in this respect Kenya is not different from almost any African country south of the Sahara — that agricultural development should rely almost totally on outside loans. Foreign aid is expected to provide not only the capital for most of the industrial expansion, and the major capital developments in agriculture, but also for services like land consolidation, extension, small loans etc. which preferably should be financed by internal revenues. As long as the beneficiaries are made to carry the full costs, as is the case with the Kenya Tea Development Authority, loans for services can probably be justified. Most of the agricultural development propositions, however, are organized in such a way that smallholders receive a heavy element of subsidy, usually in a hidden form, i.e. loan repayments are allowed to remain outstanding, settlement farmers are not charged with the full costs of personnel, irrigation settlers are not charged with the full costs of land development and water procurement, grazing development is carried out without the likelihood of the respective agencies receiving the fees necessary to cover costs. In actual practice, the policy of heavily subsidizing smallholder agriculture — which can be considered as a colonial heritage — is continued in Kenya. Rapid progress in African farming is necessary and possible, and it will be unavoidable to subsidize the introduction of innovations (where the infant industry argument applies) or nucleus estates, or settlements (where a multiplication of innovations is required), but it is doubtful whether the present amount of subsidy for smallholder farming — and almost every scheme or proposition contains an element of public support which is not recuperated — can be carried by the remaining economy for very long.

Finally it should be pointed out that sustained economic growth in Kenya — more than in most other African countries — is highly dependent on the free movement of goods and labour within a market which should preferably be even greater than the East African Common Market. Free traffic in goods is required for the expansion of industry, which in turn provides local buying power for smallholder produce. The pressing population problems could be eased if Kenya's rural surplus population could and would develop the habit of moving into expansion areas in Tanzania, Uganda and in particular in the Congo, where ample land is available. The industrialisation of Germany in the 19th century, for instance, was eased by the migration of millions of farmers to North America. And it could only be to the advantage of Africa if similar movements were to take place there. The countries in question could also benefit by an influx of Kenya farmers with a decade of experience in improved farming behind them.

G. Appendices

Appendix I: Ecological Zones of Kenya[1]

	Sq. miles
1. Afro-alpine moorland. Above 9000 feet altitude. Mainly National Parks. Low range potential.	228
2. Forests and grassland or bushlands. Potential for forestry and intensive agriculture below 7000 feet altitude. A few acres still under range use with a carrying capacity of 2—6 acres per stock unit.	19,620
3. Moist woodland, bushland or savanna. Agricultural potential good soil and topography permitting. Areas under range use still extensive and under intensive management have a carrying capacity of less than 5 acres per stock unit.	18,011
4. Dry woodland, bushland or savanna. Marginal agricultural potential. Potentially good range land — generally required less than 10 acres per stock unit. Wildlife important. Rainfall 25—35 inches.	20,601
5. Commiphora/Acacia country. Land only locally suited to agriculture in higher rainfall areas of the zone. Carrying capacity not less than 10 acres per stock unit. Perennial grasses dominant. Rainfall 25 to 12 inches.	116,436
6. Semi-desert. Bush grassland or dwarf shrub grassland. Range land of low potential. Carrying capacity over 30 acres per stock unit. Vegetation may be confined to runnels, depressions and water courses. Perennial grasses are localised within a mainly annual grassland. Wildlife important but environment precludes high populations. Rainfall below 12 inches.	44,547
Total	219,443

Appendix II

Table 34. *Prices paid to producers for principal crops and livestock products.*

	1954	1955	1956	1957	1958	1959	1960	1961	1962	1963	1964
Clean Coffee £/to	509	427	437	521	438	393	389	315	340	280	352
Tea shs/lb	2.36	2.90	2.69	3.40	3.87	3.92	4.26	4.17	4.40	4.13	4.27
Pyrethrum shs/cwt	276	293	292	288	285	303	314	261	242	271	
Sisal £/to	65	59	60	53	53	68	81	74	78	118	
Wheat shs/bag	53	52	52	53	52	52	49	47	47	47	
Maise shs/bag	39	35	35	38	35	27	32	36	36	24	
"A" Grade meat shs/lb	1.10	1.10	1.18	1.20	1.20	1.20	1.25	1.30	1.30	1.30	
Butterfat shs/lb	3.33	3.21	3.15	3.07	2.38	2.77	2.83	2.82	2.89	2.92	

Source: Statistical Abstracts: — Kenya Trade and Supplies Bulletin — Economics and Statistics Division.

[1] According to J. F. Peberdy. Ministry of Agriculture, Nairobi.

Appendix III

Table 35. *Small farm area — quantity of principal crops marketed* [1] *(1954—1964)*
'000 Tons

	1954	1955	1956	1957	1958	1959	1960	1961	1962	1963	1964 (2)
Maize	129.0	116.0	58.1	57.0	69.6	79.7	73.2	63.0	69.3	98.0	44.8
Rice (Paddy)	..	5.6	2.8	4.3	5.3	4.3	9.8	14.6	14.3	13.0	13.6
Pulses	8.0	4.5	3.9	9.2	15.5	12.5	11.3	13.0	8.7	12.4	13.7
Castor Seed.	..	..	..	..	4.6	1.8	2.7	0.7	3.6	4.5	2.4
Vegetables & Fruit	..	..	..	..	15.1	12.1	13.4	17.0	10.1	9.5	10.6
Coffee (Clean)	0.3	0.7	0.8	1.5	2.3	3.6	4.6	7.3	8.1	9.4	15.2
Tea (Manufactured)	—	—	—	—	—	0.1	0.1	0.2	0.5	0.6	0.6
Sisal Fibre & Tow	0.5	0.3	0.7	0.1	1.3	1.5	5.6	6.1	2.0	7.0	6.2
Cotton Lint	..	2.8	3.5	1.8	2.5	3.1	3.4	0.8	2.5	2.6	3.0
Wattle Bark	40.5	46.9	38.2	25.8	25.4	18.1	19.0	6.8	22.0	22.3	2.6
Cashew Nuts	1.3	1.5	2.3	2.4	1.9	3.3	5.0	6.4	2.9	4.3	5.2
Pyrethrum Flowers	0.2	0.3	0.3	0.4	0.4	0.6	1.8	2.8	2.7	1.8	2.2

[1] Marketed Production is defined as that portion of production which is sold outside the growing district.

[2] Including Resettlement Areas.

Source: Kenya: Economic Survey 1965. Nairobi 1965. See also preceding years.

Appendix IV: Some Information about the Districts of Nyeri, Elgeyo-Marakwet and Central Nyanza

1. Nyeri

Nyeri District lies just south of the equator and is bordered on the west by the Aberdare range and on the north-east by Mt. Kenya. The altitudes drop gradually towards the south-east. The population is concentrated in an area lying between roughly 4,000 and 7,500 feet. Rainfall decreases with lower altitudes and ranges from 70 inches on the slopes of the Aberdares to 35 inches in the south-east. Most of the district has deep fertile soils, mainly of vulcanic origin. Of the total area of 206,000 acres, 44 per cent is considered arable. Much of the land is too steep to cultivate, except for a portion that can be used for permanent crops, and part of the land lies so high that it is too cool for double cropping, as is normally practised in the district. Almost all the 256,000 Kikuyu in Nyeri in 1962 depended for their livelihood on small-scale farming, the number of farms and farmers amounting to 43,000, with an average acreage of 4.66 per family. The main cash activities are coffee, milk, tea, pyrethrum und potatoes.

2. Elgeyo-Marakwet

The Elgeyo-Marakwet District situated just north of the equator, covers 1,029 square miles and was inhabited in 1962 by 161,000 Elgeyo and Marakwet, belonging to the group of the Kalenjin tribes. The District stretches from the

3000 feet high Kerio Valley, over the Rift Valley escarpment, to highlands of 11,000 feet. The population and agricultural production are concentrated on a plateau of 6—9000 feet, which is covered with soil of good fertility and receives a well distributed rainfall of 40—50 inches annually. There are 10,000 farms in the district, most of them in the 10—30 acres range. The main cash activities are European potatoes, Zebu-cattle, wheat, pyrethrum, milk and wool. The main subsistence activities are maize, beans, sheep und goats.

3. Central Nyanza

The district of Central Nyanza, lying to the east and north of Lake Victoria's Kavirondo Gulf, is, after the Kikuyu districts, one of the most densely populated areas of Kenya. In 1962 the population, consisting mainly of Nilotic Waluo, numbered 639,000. The larger part of the district is taken up by the extensive Kano Plain, having an altitude of 3500—4000 feet. This ist covered chiefly by fertile but not very manageable "black-cotton" soil and receives a rather irregular rainfall of 30—40 inches per annum. Agriculture in the Kano Plain is comparatively hazardous owing to frequent droughts and floods. The smaller part of the district is a 4—5000 feet high plateau, which rises sharply from the plains and receives 50—70 inches of rain. According to the sample census of 1960/61 there are 107,400 holdings in the district, with an average size of 7.34 acres, half of which is considered to be arable. Only a minor portion of the land is cropped twice per annum. There are 272,000 head of Zebu cattle in the district.

The ecology of the area makes it unsuitable for high value cash crops like tea or pyrethrum and it is marginal for coffee, even in the higher parts. However, the conditions are good not only for millet, maize, beans and maniok which are largely grown for subsistence, but also for rice, cotton, sugar cane and bananas, particularly with irrigation. In comparison with conditions in many parts of East Africa, this is a rather attractive choice of cash crops. Hedge sisal offers an additional source of income. Most of the district is free from tsetse and much of it should be quite suitable for grade cattle.

Appendix V: Data and Assumptions Underlying the Estimates of Table 25

A. Tea Development

See: Operations and Development Plans of the Kenya Tea Development Authority, schedule 2, p. 28 and schedule 7 C, p. 35. First Year: 1963/64. In (I) the excess of payments (minus interest) over receipts have been considered as investments and the production of tea, valued at Shs 1/80 per lbs dry tea (value of green leaf) as return. In (II) the excess of payments (minus interest) over receipts have been considered as investments and the excess of receipts over payments, plus interest, plus 12 per cent of the value of the green leaf produced as return. In both cases the tea produced prior to 1963/64 has been included in the estimate.

B. Agricultural Administration and Extension in Three Districts

Nyeri District. See Table 6 and 7. First Year 1953. £ 20,000 have been considered as administration costs and all expenditure over £ 20,000 as investment. Expenditure for famine relief work in 1954 has not been included. The marketed production of 1953—1958 has been considered as the base. In (I) the additional marketed production thereafter has been related to the investments. In (II) 12 per cent of the value of the marketed production, plus — since 1964 only — the export tax on coffee, has been related to the investments. The estimate is based on the assumption that expenditure and marketed production after 1963 or 1964 remain at the 1963 and 1964 level.

Elgeyo-Marakwet District. See Table 6 and 7. First Year 1953. All expenditure over £ 9,000 has been considered as investments. A marketed production of £ 100,000 has been considered as the base. In (I) the additional marketed production after 1958 has been related to the investment. In (II) 12 per cent of the value of the marketed production has been related to the investments. The estimate is based on the assumption that expenditure remains at the 1963 level and marketed production on the average of 1963 and 1964.

Central Nyanza District. See Table 6 and 7. First Year 1953. All expenditure over £ 20,000 has been considered as investment. The marketed production 1954—1958 has been considered as the base. In (I) the additional marketed production thereafter has been related to the investment. In (II) 12 per cent of the value of the marketed production have been related to the investment. The estimate is based on the assumption that expenditure after 1963 remains at the 1963 level and that the marketed production remains on the average of 1963 and 1964. The relatively high return in terms of marketed production is entirely due to this assumption which is an optimistic view in the case of Central Nyanza.

C. Irrigation Development

Investments see Table 12. The estimate is based on annual operational savings in Mwea-Tebere of £ 50,000 and no operational savings or losses at Perkerra from 1964/65 on. In (I) the investment is related to an annual production value of £ 265,000 at Mwea-Tebere and £ 50,000 at Perkerra. For the earlier years of the schemes correspondingly lower figures have been used. In (II) the investment is related to 12 per cent of the value of the marketed production, plus operational savings, plus £ 25,000 for the rice stabilisation fund (for Mwea-Tebere only).

D. Resettlement in Low Density Schemes

The assumptions are the following: Total investment in resettlement: £ 27 per acre. The settler is expected to reach the marketed production of £ 8,1 per acre. Before resettlement marketed production has been £ 4.7 per acre. The settler is expected to pay the following instalments per annum: £ 2.1 per acre in the first 10 years and £ 0.9 thereafter. The calculations show the rate of return in case production and loan repayment is reaching 90 per cent and 70 per cent of these targets.

E. New Settlement on Rain-fed Land

The assumptions are the following: Total investment: £ 150 per family. No operational savings and losses. The (additional) marketed production per family is £ 20 per annum. See Table 11.

In all five cases it has been assumed that all additional marketed production is increasing the revenues of the Treasury, by direct taxes on goods bought by the farmer, at the rate of 12 per cent of the sales.

Bibliography

(Literature concerning the economics and sociology of African farming in Kenya)

AMIR, A.: Kenya's Problems of Agricultural Settlement and Development. Nairobi 1962, Mimeo.

APTHORPE, R. J.: (1) Sociology of Settlement Schemes. (2) Sociological Aspects of Development Planning for Rural Areas in East Africa. Makerere Univ. Coll. (in prep.).

ASKWITH, T. G.: Progress through Self Help. Nairobi 1960.

BAKER, S. J. K.: Pastoralist and Cultivator in the Highlands of East Africa. Int. Congr. of Anthr. and Ethn. Kopenhagen 1939.

BARWELL, C. W.: A Note of Some Changes in the Economy of the Kipsigis Tribe: J. Afr. Adm. Vol. XIII, No. 2, 1956.

BARBER, W.: Some Questions about Labour Force Analysis in Agrarian Economics with Particular Reference to Kenya. CER Discussion Paper No. 13, Nairobi 1965.

BARWELL, C. W.: Farmers' Training Centres. Min. of Agr. Nairobi, Mimeo.

BELSHAW, D. G. R.: Agricultural Settlement Schemes in the Kenya Highlands. E. A. Geogr. Rev. No. 2, April 1964.

— Agricultural Production and Trade in the East African Common Market. Makerere 1964, Mimeo.

BISSCHOP, H. R.: The Improvement of Livestock in Kenya. Min. of Agr. Nairobi 1949, Mimeo.

BRADFORD, E. L.: African Mixed Farming Economics as Applied to Bukura. E. A. A. J. Vol. XII, July/Oct. 1946.

— Bukura Native Agricultural School. E. A. A. J. Vol. XI, Jan./April 1946.

BRANAGAN, D.: A Discussion on the Factors Involved in the Development of Masailand. Nairobi 1962, Mimeo.

BRETT, E. A.: Economic Policy in Kenya Colony: A Study in the Politics of Resource Allocation. EAISR Conf. 1965, Mimeo.

BRIDGER, G. A.: Planning Land Settlement Schemes: ECA-FAO. Agr. Econ. Bull. for Africa No. 1, 1962.

BROUGH, A. T.: Comment on the Report of the Committee on the Organisation of Agriculture. E. A. Econ. Rev. Vol. 7, No. 2, 1960.

BROWN, F. A.: Report on Irrigation Schemes. Min. of Agr. Nairobi 1955, Mimeo.

— A Review of the Perkerra River Scheme. Min. of Agr. Nairobi 1954, Mimeo.

— A Review of the Mwea Tebere Irrigation Scheme. Min. of Agr. Nairobi 1955, Mimeo.

BROWN, L. H.: A National Cash Crop Policy for Kenya. Part I & II. Nairobi 1963.

— Land Consolidation and Better Farming in Kenya. Emp. J. Exp. Agr. Vol. 30, No. 120, 1962.

— Draft Agricultural Policy, Nyanza Province. Min. of Agr. 1954, Mimeo.

— Revised and Consolidated Agricultural Policy, Central Province. Min. of Agr. 1952, Mimeo.

— A Survey of the Grazing Schemes Operating in Kenya. Min. of Agr. 1959, Mimeo.

— Development and Farm Planning in the African Areas of Kenya. E. A. A. J. Vol. XXIII, No. 2, Oct. 1958.

Brown, L. H.: The Development of the Semi-Arid Areas of Kenya. Min. of Agr. 1964, Mimeo.
Brown, P. H.: Stock Reduction in Nandi: J. Afr. Adm. Vol. X, No. 1, Jan. 1958.
Bullock, R. A.: Landscape Change in Kiambu. E. A. Geogr. Rev. No. 3, April 1965.
Burgwin, W. A.: Smallholder Sisal Production. Kenya Sisal Board Bulletin, July 1964.
Byant-Mumford, W.: East Africa, Some Problems in Native Development and a Possible Solution in Co-operative Societies. Africa, Vol. VI, 1933.
Carey-Jones, N. S.: Why Organise in Agriculture? E. A. Econ. Rev. Vol. 7, No. 2, 1960.
Carson, J. B.: Some Experience in Native Agriculture in the Elgeyo-Marakwet District of Kenya. E. A. A. J. Vol. XIII, July 1947.
Cavendish-Bentick, F. W.: Agricultural Policy in African Areas of Kenya. Nairobi 1951.
Chambers, P. C.: Planned Group-Farming in Nyanza Province, Kenya. UN Econ. Soc. Coun. 1949, Mimeo.
— Group Farming in Kenya. Corona No. 2, July 1950.
Chaundy, G. H.: The Agricultural Education of a Primitive Tribe. The West Suk of Kenya. E. A. A. J. Vol. XIII, April 1943.
— Primitive Agricultural Methods of the West Suk Tribe and Improvements. E. A. A. J. Vol. V, July 1939.
Clayton, E. S.: Policies Affecting Agrarian Development in Kenya. E. A. Econ. Rev. Vol. 5, No. 2, 1959.
— Safeguarding Agrarian Development in Kenya. J. Afr. Adm. Vol. XI, No. 3, July 1959.
— Economic and Technical Optima in Peasant Agriculture. J. Agr. Econ., May 1961.
— Peasant Coffee in Kenya. World Crops. Vol. 13, July 1961.
— Small Scale Cash Crop Production in a Developing Economy. Econ. Dev. and Cult. Change Vol. 9, No. 4, July 1961.
— A Note on the Economics of African and European Land Use in Kenya. E. A. Econ. Rev. Vol. 8, No. 1, June 1961.
— Cash Crops for Smallholders. World Crops, Vol. 13, No. 8, Aug. 1961.
— A Note on the Alien Enclave and Development. E. A. Econ. Rev. Vol. 10, No. 1, 1963.
— Agrarian Development in Peasant Economies: Some Lessons from Kenya. London 1964.
Clough, R. H.: E. A. Agricultural Diploma Manpower Survey. Egerton 1965, Mimeo.
— Economic Survey of Land Settlement, Mau Tuma Scheme (Lugari). Egerton 1964, Mimeo.
— Some Economic Aspects of Land Settlement in Kenya. E. A. Ag. Econ. Conf. Nairobi 1965, Mimeo.
— Some Economic Aspects of Land Settlement in Kenya. Egerton 1965, Mimeo.
Colonial Office: East African Royal Commission 1953—1955 Report. London 1955, Cmd. 9475.
— Land and Population in Kenya. H. M. S. O. No. 290, London 1952.
— Some Notes on Agricultural Develpoment Schemes in British Colonial Territories. London 1955, Mimeo.
— Comments on Royal Commission Report. London 1932.
— Memorandum on Native Policy. London 1930.

COWLEY, G. S.: An Outline of the Technical Revolution in Agricultural Production in Kenya 1946—1956. Min. of Agr. Nairobi, Mimeo.

CCTA: Conference on the Mechanisation of Agriculture. Entebbe 1955, Mimeo.

CRIDDLE, D. W.: Irrigation Development and Practices of Kenya. Nairobi 1961, Mimeo.

DAKEYNE, R. B.: The Pattern of Settlement in Central Nyanza. Australian Geogr. Vol. VII, No. 4, March 1962.

DAVIDSON, B. R.: The Economics of Arable Land Use in African and European Areas of Kenya. E. A. Econ. Rev. Vol. 7, No. 1, June 1960.

—, and R. J. YATES: The Potential Supply of Arable Land in Kenya and its Productivity. E. A. Econ. Rev. Vol. 7, No. 1, 1960.

— — Relationship Between Population and Potential Arable Land in the African Reserves and the European Highlands. E. A. Econ. Rev. Vol. 6, No. 2, Dec. 1959.

DESHLER, W.: Land Use by the Bajun People of the Northern Kenya Coast. EAISR 1954, MS.

DUNN, R. P.: Cotton in British East Africa. Memphis, Tennessee, 1949.

E. A. High Commission: Report on the Analysis of the Sample Survey of African Agriculture. Nairobi 1950/51.

East African Royal Commission and African Land Tenure: J. Afr. Adm. Vol. VIII, No. 2, April 1956.

EDEN, T.: Report on a Visit to African Tea Smallholdings in Kenya, 1957, Mimeo.

EDWARDS, D. C.: Report on the Grazing Areas of the Northern Frontier District. Min. of. Agr. Nairobi 1944, Mimeo.

ENLOW, C. R.: The Grassland Situation in Kenya. Min. of Agr. 1958, Mimeo.

— Some Observations on Agriculture in Kenya, 1958. Nairobi 1961.

ETHERINGTON, D. M.: Land Re-Settlement in Kenya: Policy and Practice. E. A. Econ. Rev. No. 1, 1963.

— Projected Changes in Urban and Rural Population in Kenya and the Implications for Development Policy. Univ. Coll. Nairobi 1965, Mimeo.

— The Economics of Peasant Tea Production in Kenya. Univ. Coll. Nairobi (in prep.).

Fabian Colonial Bureau: Opportunity in Kenya. No. 162. London 1953.

FALLON, E. L.: Famine Relief, Kajiado District. Nairobi 1962, Mimeo.

— Turkana Grazing Control Scheme. Min. of Agr. (?).

Farm Economic Survey Unit: Economic Investigation of Farming in Nyeri District, 1963—64. Preliminary. Nyeri 1964, Mimeo.

FAULKNER, D. E.: The Development of the Livestock Industry of the Native Areas of Kenya, Policy and Plans 1949—1958. Nairobi 1948.

— Notes on Animal Health and Industry for Africans. Min. of Agr. Nairobi 1953, Mimeo.

FEARN, H.: Population as a Factor in Land Usage in the Nyanza Province of Kenya Colony. E. A. A. J. Vol. 20, Jan. 1955.

— Cotton Production in the Nyanza Province of Kenya Colony, 1908—1954. Emp. Cot. Grow. Rev. Vol. XXXIII, No. 2, 1956.

— An African Economy. A Study of the Economic Development of the Nyanza Province of Kenya, 1903—1953. London 1961.

FISHER, J.: The Anatomy of Kikuyu Domesticity and Husbandry. Nairobi 1952, Mimeo.

FLIEDNER, H.: Some Legal Aspects of Land Reform in Kenya. EAISR Conf. 1963, Mimeo.

Fliedner, H.: Die Bodenrechtsreform in Kenya, Heft 7, Afrika-Studien des Ifo-Institutes, Berlin — Heidelberg — New York 1965.

FAO: Seminar on Agricultural Education, Kampala 1963. Rome 1963.

— Reconnaissance of the Agricultural Potential of the Turkana District of Kenya. Rome 1964, Mimeo.

— Report of the Grassland Missions in Kenya. Rome 1955, Mimeo.

— Joint FAO/UNICEF Report on the Marketing of Milk in the Coastal Region of Kenya. Rome 1960.

— Milk Development Possibilities in East Africa. Rome, June 1958.

Forrester, M. W.: Kenya To-day. s'Gravenhage, 1962.

Gamble, G.: Report on Visits to India, Malaya and Ceylon with some Notes for the Guidance of Tea Planters in Kenya. Nairobi 1951.

Gerlach, P. I.: Economy and Malnutrition among the Digo. EAISR 1961. MS.

Giglioli, E. G.: Mechanical Cultivation on the Mwea Irrigation Settlement. E. A. A. J. Vol. XXX, No. 3, Jan. 1965.

— Staff Organisation and Tenant Discipline on an Irrigated Land Settlement. E. A. A. J. Vol. XXX, No. 3, Jan. 1965.

Golding, F. R. H.: African Coffee. Coffee Board of Kenya. Nairobi.

Graham, M. D.: An Experiment in Native Mixed Farming in the Nyanza Province of Kenya. E. A. A. J. Vol. VIII, Oct. 1941.

Green, E. C., A. T. Brough, and W. Wamala: Report of the Working Party Set up to Examine the Cashew Nut Industry in Kenya. Nairobi 1963, Mimeo.

Gregory, R. G.: Sidney Webb and East Africa. Berkley and Los Angeles 1962.

Grimes, R. C.: Crop Husbandry for the Coastal Areas. Min. of Agr. Mimeo.

Guillard, J.: Conservation Measures for African Soils: High Altitude Native Areas in Kenya. Sols Africaines. Vol. 2, No. 3/4, 1953.

Gulliver, P.: A Preliminary Survey of the Turkana. Cape Town 1951.

— The Family Herds of two Pastoral Tribes in East Africa, the Jie and the Turkana. London 1957.

Hailey: Native Administration in British African Territories. London 1956.

Hall, A. D.: The Improvement of Native Agriculture in Relation to Population and Public Health. London 1936.

Harrison, E.: Native Agriculture in Kenya. Nairobi 1926.

Heady, H.: Range Management in East Africa. Nairobi 1960.

Henderson, G. R., and P. T. Preston: Fodder Farming. Min. of Agr. Nairobi.

—, and G. N. Williamson: The Field Unit Size System for Importing Flexibility to Farm Layouts and Plans for Smallholdings. Nyeri 1959.

Hennings, R. O.: Grazing Management in the Pastoral Areas of Kenya. J. Afr. Adm. Vol. XIII, No. 4, Oct. 1961.

— Some Trends and Problems of African Land Tenure in Kenya. J. Afr. Adm. Vol. VI, No. 4, Oct. 1952.

Herskovits, M. J.: The Cattle Complex in East Africa. American Anthropologist Vol. 28, 1926.

Heyer, J. U.: An Investigation of the Limits on Peasant Agricultural Production. Case Studies from Machakos and Kericho Districts of Kenya. Univ. Coll. Nairobi (in prep.).

— Seasonal Labour Inputs in Peasant Agriculture. Univ. Coll. Nairobi 1965, Mimeo.

— Linear Programming Model for Peasant Agriculture in Kenya. EAISR Conf. 1965, Mimeo.

Hill, M. F.: The Dual Policy in Kenya. London 1943.

History and Customs Relative to the Settlement of Kiambu. Min. of Agr. Nairobi 1957. MS.

Homan, F. D.: Inheritance in Kenya Native Land Units. J. Afr. Adm. Vol. X, No. 3, July 1958.

— Land Consolidation. J. Afr. Adm. Jan. 1961.

— Succession to Registered Land in the African Areas of Kenya. J. Local Adm. Overseas Vol. II, No. 1, Jan. 1963.

— Land Consolidation and Redistribution of Population in the Imenti Sub-Tribe of the Meru, Kenya. African Agrarian Systems. Oxford 1963.

—, and R. A. Sands: Land Tenure Reform and Agricultural Development in the African Lands of Kenya. Nairobi, Mimeo.

Hughes, O. E. B.: Village in Kikuyu Country. J. Afr. Adm. Vol. VII, No. 4, Oct. 1955.

Hughes-Rice, J.: Soil Conservation Organisation in Fort Hall District. E. A. A. J. Vol. XII, Jan.—April 1947.

Hughes-Rice, T.: Tea Development in Nyanza African Areas. Tea Vol. 1, No. 4, 1960.

Humphrey, N.: The Gede Native Settlement Scheme. E. A. A. J. Vol. IV, May 1939.

— The Liguru and the Land. Nairobi 1947.

— Thoughts on the Foundations of Future Prosperity in the Kikuyu Lands Based on a Review of some Factors Affecting Agriculture in the South Nyeri District. Min. of Agr. 1949, Mimeo.

— A Preliminary Report on Agricultural Conditions in South Nyeri. Nyeri 1944, Mimeo.

—, E. H. Lambert, and P. W. Harris: The Kikuyu Lands. Nairobi 1945.

Huntingford, G. W. B.: The Nandi Work and Culture. London 1950.

— The Economic Life of the Dorobo. Anthropos. Vol. 50, No. 4/6, 1955.

Huxley, E.: A New Earth. London 1961.

IBRD: Notes of the World Bank Mission to Kenya, 1961, Memoranda. Nairobi, Mimeo.

— The Economic Development of Kenya. Baltimore 1963.

James, L.: The Kenya Masai, a Nomadic People under Modern Administration. Africa Vol. XIII, Jan. 1939.

Johnston, B. F.: The Choice of Measures for Increasing Agricultural Productivity: A Survey of Possibilities in East Africa. Trop. Agr. Vol. 41, No. 2, 1964.

Jones, P. H.: The Marketing of African Livestock. Nairobi 1959, Mimeo.

Jordan, S. M.: Reclamation and Pasture Management in the Semi-Arid Areas of Kitui District, Kenya. E. A. A. J. Vol. XXIII, No. 2, Oct. 1957.

Joy, J. L.: Some Agricultural Aspects of the Report (Royal Commission). E. A. Econ. Rev. Vol. 2, No. 2, Jan. 1958.

Juckes, D.: Land Use in Elgon Nyanza District. Min. of Agr. Nairobi, MS.

Karue, Ch. H.: Peasant Coffee Production in Mathira, Kenya. Makerere Univ. Coll., MS.

Kennedy, T. J.: A Study of Economic Motivation Involved in Peasant Cultivation of Cotton. E. A. Econ. Rev. Vol. 10, No. 2, 1962.

— Farm Economic Survey Nyanza and Western Regions. Nakuru 1964, MS.

Kenya: Accelerated Development of African Agriculture, 1958/63. Min. of Agr., Mimeo.

— African Land Development Organisation (Aldev) Annual Reports. Nairobi.

— African Development in Kenya, 1945/55. Land Livestock and Water. Nairobi 1956.

Kenya: African Land Development in Kenya. Nairobi 1962.
— Agricultural Policy and Development Proposals, 1964—1970. 1964, Mimeo (for each region).
— Agricultural Policies in African Areas. Nairobi 1951.
— (1) Agricultural Problems and Potentials of the African Land. (2) Progress of Group Farm. (3) Report on the Agrarian Policy. Memoranda to Royal Commission Report. Nairobi 1953, Mimeo.
— Agricultural Review. Treasury 1962, Mimeo.
— Annual Reports of the Development and Reconstruction Authority. Nairobi (up to 1953).
— Annual Review. Dep. of Agr. Nairobi.
— Better Farming in Central Province. Nairobi 1962.
— Brief Summary of a Scheme for the Development of some 200,000 Acres of the Masai Land Unit in Kenya for the Production of Wheat and Wooled Sheep. Min. of Agr. 1965, Mimeo.
— Communique on Land Tenure Policy. Nairobi 1951.
— Development Plan, 1964—1970. Nairobi.
— Department of Settlement, Annual Reports. Nairobi.
— Directive on the Granting Loans to Africans for Agricultural Purposes. Min. of Agr. 1952, Mimeo.
— Ecology as a Basis of Agricultural Development. Min. of Agr. 1953, Mimeo.
— Economic Survey of the Dairy Industry in Kenya. Econ. Stat. Div. 1963, Mimeo.
— Experimental Smallholding of the Scotts Agricultural Laboratories. Nairobi 1932.
— Expansion of Tea Planting in the Central Province. Nyeri 1958, Mimeo.
— The Family Farm. Nairobi 1933.
— Final Report of the Sisal Working Party. Min. of Agr. 1962, Mimeo.
— Future of the Tea Industry, Memorandum. Min. of Agr. 1947, Mimeo.
— General Aspects of the Agrarian Situation in Kenya. Nairobi 1946.
— Grassland and Forage Farming in the High Bracken, Kikuyu, and Star Grass Zone of Mt. Kenya and the Eastern Aberdares. Min. of Agr. Nairobi 1963, Mimeo.
— History of the Fluctuation of the Cattle Population, in Kajiado Masailand during the Last Twenty Years. Min. of Agr. 1962, MS.
— House of Representatives. Official Reports. Nairobi Cont. (see various debates about agricultural policy).
— Kenya African Agricultural Sample Census, 1961/62. Part I & II. Nairobi.
— Land Development Board. Annual Reports. Nairobi.
— Land and Agricultural Bank of Kenya. Annual Reports. Nairobi.
— Land Utilisation and Settlement. Sessional Paper No. 8. Nairobi 1945.
— Land Utilisation and Settlement: A Statement of Government Policy. Nairobi 1945.
— Memorandum on Agricultural Development in the Native Reserves. Nairobi 1957, Mimeo.
— Native Land Tenure in Kikuyu Province. Nairobi 1929.
— Native Land Trust Ordinance. Nairobi 1949 (Chapter 100).
— Ordinance to Promote and Maintain a Stable Agriculture. Nairobi 1955.
— Progress in African Agriculture. E. A. A. J. Vol. XVIII, No. 2, Oct. 1951.
— Reform of Land Tenure, the Development of Agriculture and the Use of Farm Surveys in the African Peasant Farming Areas of Kenya. FAO Seminar, Nairobi 1961, Mimeo.

Kenya: Relationship of Farm Institute Training to Extension — Extension and Follow-up Methods in African Areas of Kenya. In: FAO/CCTA, Arusha 1962, Mimeo.
— Reports of the Coconut Commission. Nairobi 1914.
— Mariakani Milk Scheme. Debates. Nairobi, Dec. 1960.
— Report of the Committee on Agricultural Credit for Africans. Nairobi 1950.
— Report of the Committee to Carry out an Economic Survey of South Nyanza and Kericho District. Nairobi 1957, Mimeo.
— Report on Famine Relief in Kenya. Nairobi 1962.
— Report of the Kenya Land Commission. Nairobi 1933.
— Report of the Working Party on African Land Tenure, 1957—58. Nairobi 1958.
— Reports of the Working Party Set Up to Consider the Establishment of an Authority to Promote the Development of Cash Crops for Smallholders and of the Working Party Set up to Consider the Financial Implications of the Proposed Authority. Nairobi 1960 (?).
— Review of Agriculture. Econ. Stat. Div. 1963, Mimeo.
— Review of the Organisation and Administrative Activities of the Ministry of Agriculture's Loans to Farmers. Min. of Agr. 1961, Mimeo.
— Revised Edition of the Agricultural Ordinance 1955. Nairobi 1961.
— Revised and Consolidated Agricultural Policy. Central Province. Min. of Agr. 1953, Mimeo.
— Three Year Report, 1955—57, 1958—60. Min. of Agr. Nairobi.

Kenya African Democratic Union: Land Tenure and Agricultural and Pastoral Development for Independent Kenya. Nairobi 1962.

Kenya Central Land Board: Annual Reports. Nairobi.

Kenya Land Commission: Evidence and Report. Nairobi 1933.

Kenya Tea Development Authority: The Operations and Development Plans. Nairobi 1964.

Kershaw, G.: Social and Economic Consequences of Land Consolidation in Kiambu District. EAISR Mimeo.

Kiambati, J. M.: Rural Re-Settlement in Kenya. Meeting on Rural Settlement. Israel 1965, Mimeo.

Kolbe, L. H., and S. J. Fouche: Land Consolidation and Farm Planning in the Central Province. Nairobi 1959.

Lambert, H. E.: The System of Land Tenure in the Kikuyu Land Unit. Cape Town 1958.

Lambert, N.: Land Tenure Among the Akamba. Afr. Stud. Vol. 6, No. 3, 1947.

Leaky, L. S. B.: The Economics of Kikuyu Tribal Life. E. A. Econ. Rev. Vol. 3, No. 1, July 1956.

Lindsell, M.: Legal Aspects of African Large Farms. Unpublished Manuscript. Reg. Agr. Off. Nakuru 1964.

Lynn, C. W.: Agricultural Extensions and Advisory Work with Special Reference to the Colonies. London 1949, Mimeo.

MacArthur, J. D.: Land Tenure Reform and Economic Research into African Farming in Kenya. E. A. Econ. Rev. Vol. 8, No. 2, 1961.
— The Development of Research into the Production Economics of African Peasant Farms in Kenya. E. A. Econ. Rev. Vol. 9, No. 2, Dec. 1962.
— Some Economic Aspects of Agricultural Development in Nyeri District 1962. Farm Economic Survey Unit. Nakuru 1964, Mimeo.
— Some Economic Case Studies of Farms in Nandi and West Pokot District, 1963—64. Farm Economic Survey Unit. Rep. Nr. 23. Nakuru 1965, MS.

MacArthur, J. D.: The Analysis and Use of Farm Management Data in Kenya. E. A. Agr. Conf. 1965, Mimeo.
— Economic Aspects of Land Settlement in Kenya (in prep.) (see also detailed manuscripts over Aldev schemes).
— Some Economic Studies of African Farms in Rift Valley Province 1961. Farm Economic Survey Unit. Nakuru 1963.
— Some Economic Case Studies of Farms in Nandi District, 1962—63. Farm Economic Survey Unit. Nakuru 1964.
— Some Economic Studies of Farms in Elgeyo and West Pokot Districts, 1962. Farm Economic Survey Unit. Nakuru 1963.
MacGillivray, D., A. A. Lawrie, and H. White: Report of the Committee on the Organisation of Agriculture. Nairobi 1960.
Maher, C.: Soil Erosion and Land Utilisation in the Ukamba Reserve. Min. of Agr. Nairobi 1937.
— Soil Erosion and Land Utilisation in the Kamasia, Njemps and East Suk Reserves. Min. of Agr. Nairobi 1937.
— Soil Erosion and Land Utilisation in the Embu Reserve. Min. of Agr. Nairobi 1938.
— Notes on Soil Erosion and Land Utilisation in Nyanza Province. Min. of Agr. Nairobi 1938.
— The People and the Land. Part I and II. In: E. A. A. J. Vol. 7, No. 2 & 3, 1941/42.
— The Development of Soil Conservation in Kenya Colony. Part I and II. World Crops Vol. 3, June-July 1951.
Mann, I.: The Conversion of Unproductive Animals from Overstocked Areas into High Protein Human Food. Kabete 1959, Mimeo.
Martin, C. J.: Report on the Analysis of the Sample Survey of African Agriculture. E. A. Stat. Dep. Nairobi 1950.
Masefield, G. B.: A Comparison Between Settlement in Villages and Isolated Homesteads. J. Afr. Adm. Vol. X, 1957.
— Group Farming with Special Reference to East Africa. World Crops Vol. 2, March 1950.
Matheson, J. K., and E. W. Bovill: East African Agriculture. London.
Maunder, A. H.: Agricultural Extension in Kenya. Min. of. Agr. Nairobi 1962, Mimeo.
Mayer, P.: Agricultural Co-operation by Neighbourhood Groups among the Gusii. Col. Res. Nr. 3. London 1951.
Mburu, J. S.: Agriculture in the Kiambu Reserves. Min. of Agr. 1956 (?), Mimeo.
McCulloch, A. F.: Animal Husbandry in the African Areas of Kenya. Min. of Agr., Mimeo.
McEntee, P. D.: Improved Farming in the Central Nyanza District. J. Afr. Adm. Vol. XII, No. 2, April 1960.
McGlashan, N. D.: Resettlement in the Meru-District of Kenya. Geography Vol. 43, July 1958.
McLoughin, P. F. M.: The World Bank Study of African Agriculture, some initial Aspects. Washington 1963, Mimeo.
McWilliam, M. D.: The East African Tea Industry. Part V, Tea Grown by Africans. Oriel and Nuffield Colleges. Oxford 1957, Mimeo.
McWilliam, M. W.: The Kenya Tea Industry. E. A. Econ. Rev. Vol. 6, No. 1, 1959.
Melville, A. R.: Introduction of Exotic Cattle into Peasant Farming in Kenya. Conf. Wye College, Spt. 1961, Dep. of Techn. Coop. Misc. No. 2.

Middleton, J.: The Kikuyu and Kamba of Kenya. London 1953.
Mitchell, P. E.: Land and Population in East Africa. London 1952.
— The Agrarian Problem in Kenya, Nairobi 1947.
Neviger, N. T.: New Forms and Developments of Joint Farming and Agricultural Co-operation in East Africa. Ifo-Institut, Munich (in prep.).
Njukia, J.: Smallholder Tea Production in Mathira Division of Nyeri District 1964. Makerere Univ. Coll., MS.
Njugana was Gakuo: Bedeutung und Möglichkeiten des Genossenschaftswesens für die Entwicklung der Wirtschaft Kenyas. Diss. Freiburg 1960.
Oliver, S. C.: Individuality, Freedom of Choice, and Cultural Flexibility of the Kamba. American Anthropologist Vol. 67, No. 2, 1965.
Omamo, W. O.: Mechanisation of Saga Group Farms in the Lake Victoria Foreshore Location. Min. of Agr. 1965, Mimeo.
— A Comparative Economic Study of Consolidated Holdings in Nyeri District in the Central Province of Kenya Colony. M. Sc. Thesis, Univ. of the Punjab 1959.
Ominde, S. H.: Some Aspects of Land and Population Problems in the Lake Districts of Western Kenya. E. A. Academy Symposium 1963, Mimeo.
Orchardson, T. Q.: Future Development of the Kipsigis with Special Reference to Land Tenure. J. E. Afr. Ug. Nat. Hist. Soc. Vol. 12, No. 5/6, 1935.
— The Kipsigis. 1961.
Payne, W. J. A.: Relation of Animal Husbandry to Human Nutritional Needs in East Africa. E. A. A. J. Vol. 29, 1963—64.
Peberdy, J. F.: A Study of Some Aspects of the Machakos District Economy. Min. of Agr. Nairobi 1961, Mimeo.
Pedraza, C. J. W.: Land Consolidation in the Kikuyu Areas of Kenya. J. Afr. Adm. Vol. VIII, No. 2, April 1956.
Peers, A. W.: The Agriculture of the Eweso Ng'iro Basin. An Investigation into the Water Resources of the Eweso Ng'iro Basin. Rep. No. 4, Vol. II, Min. of Works. Nairobi 1962, Mimeo.
Penwill, D. J.: A Pilot Scheme for two Kikuyu Improved Villages. J. Afr. Adm. Vol. XII, No. 2, April 1960.
— Provisional Development Policy, Kiambu District. Min. of. Agr. Nairobi, Mimeo.
Peristiany, J. G.: The Social Institutions of the Kipsigis. London 1964.
Pilgrim, J. W.: Landownership in the Kipsigis Reserve. EAISR Conf. 1965, Mimeo.
Planned Group Farming in Nyanza Province, Kenya. Trop. Agr. Vol. 27, No. 7 to 12, July/Dec. 1950.
Pollock, N. C.: Agrarian Revolution in Kikuyuland. South Afr. Geogr. J. Vol. 41, Dec. 1959.
Porter, Ph. W.: Environmental Potentials and Economic Opportunities. — A Background for Cultural Adoption. American Anthropologist Vol. 67, No. 2, 1965.
Pratt, D. J.: The Grazing Areas of the Lower Eweso Ng'iro Basin. An Investigation into the Water Resources of the Eweso Ng'iro Basin. Rep. No. 4, Vol. II, Min. of Works. Nairobi 1962.
— Methods for Rehabilitation and Management of Rangelands in Kenya. Conf. on Rangeland. Kitale 1964, Mimeo.
Preston, P. T.: Livestock and Meat Marketing in Kenya. Fort Lamy Conf. 1960, Mimeo.
Progress in African Agriculture, an Outline of Postwar Improvements in Kenya. E. A. A. J., Vol. 18.

Report of the Kenya Working Party on African Land Tenure. J. Afr. Adm., Vol. XI, No. 4, Oct. 1959.
REID, D. R.: Preliminary Survey of Livestock Development Possibilities in East Africa. FAO Rome 1964, Mimeo.
Report of the Mission Investigating Methods of Processing Smallholder Tea Production in Central Province, Kenya Nairobi 1959, Mimeo.
RICE, J. H.: Soil Conservation Organisation in Fort Hall District. E. A. A. J., Vol. 12, April 1947.
ROOEK, J. D. DE: Social Economische Aanpassingsprobleme in Kenya. Haarlem, 1950.
RYAN, T. C. I.: A Rejoinder to Mr. Clayton's Note on the Alien Enclave and Development. E. A. Econ. Rev., Vol. 10, No. 1, 1963.
SAVILE, A. H.: Notes on the Best "Bush School" Techniques for Adult Instruction. Mombasa 1958, Mimeo.
—, and T. HUGHES-RICE: Extension Methods in African Areas of Kenya. Conf. at Wye College. Col. Off. No. 531. London 1958.
SCHNEIDER, H. K.: The Subsistence Role of Cattle Among the Pokot and in East Africa. Americ. Anthr. Vol. 59, No. 2, April 1957.
— Pokot Resistance to Change. Bacom & Herskovits, eds.: Continuity and Change in African Cultures. Chicago 1962.
SILLITOE, K. K.: Preliminary Notes of Land Tenure and Usage in Meru District. EAISR, Kampala 1962, Mimeo.
— Land Use and Community in Nyeri, Kenya. EAISR, Kampala 1963.
— Social and Economic Consequences of Land Consolidation in Nyeri District. EAISR, Kampala 1964, Mimeo.
SLUITER, G.: Report on Kikuyu Resettlement 1960. EAISR Kampala 1964, MS.
SORRENSON, M. P. K.: Counter Revolution to Mau Mau: Land Consolidation in Kikuyuland 1952—1960. EAISR, Kampala 1962, Mimeo.
— The Official Mind and Kikuyu Land Tenure, 1895—1939. EAISR, Kampala 1963, Mimeo.
— Land Consolidation and Registration in Kikuyuland. EAISR, Kampala 1964, MS.
Special Crops Development Authority: Annual Reports, Nairobi.
SPENCER, P.: The Samburu. London 1964.
STORRAR, A.: The Principles of Farm Planning. Trop. Agr. Vol. 36, No. 3, July 1959.
— A Guide to the Principles and Practices of Land Settlement in Kenya. J. Loc. Adm. Overseas, Vol. III, No. 1, Jan. 1964.
—, and D. R. BREWIN: Extension Personnel — Selection, Training Conditions of Service and Equipment. FAO/CCTA Conf. Arusha 1962, Mimeo.
SWYNNERTON, R. J. M.: A Plan to Intensify the Development of African Agriculture, Nairobi 1954.
— Kenya's Agricultural Planning. Afr. Aff., Vol. 56, July 1957.
— Planning African Farming in Kenya. Conf. Wye Coll. Col. Off., No. 531, London 1958.
— Agricultural Advances in Eastern Africa. Afr. Aff., Vol. 61, No. 244, July 1962.
SWYTEK, W. L.: Social Factors in Luo Land Consolidation. EAISR Conf. 1965, Mimeo.
THOMPSON, A. W.: African Agriculture in Kenya. East African Agriculture, London 1950.
WAGNER, G.: The Bantu of North-Kavirondo. Vol. II, Economic Life, London 1956.

WAGNER, G.: Die moderne Entwicklung der Landwirtschaft bei den Kavirondo-Bantu. Z. Ges. Erdk. 7/8, 1940.

WEBSTER, C. C.: Land Utilisation and Soil Fertility Maintenance in Kenya. Proc. CCTTA Conf. Leopoldville 1954.

WHISSON, M. G.: Economic and Social Change among the Kenya Luo.

WIJEYEWARDEBE, G. E. T.: Some Aspects of Village Solidarity in Ki-Swahili Speaking Coastal Communities of Kenya and Tanganyika. EAISR 1961, Mimeo.

WILBRANDT, H.: Kenia — ein Brandherd? In: Brandherde der Weltpolitik. Otto-Suhr-Inst., Berlin 1963.

WILDE, J. C. DE: Development of African Agriculture and Animal Husbandry in Kenya. IBRD, Washington 1964, Mimeo.

WILLIAMS, M. H.: Agricultural Credit in Kenya. Min. of Agr. 1960.

WILSON, G. M.: Village Surveys, Bunyoro, Nyangori, Maragoli, Tikiri, Boholo. Marco Surveys. Nairobi 1956, Mimeo.

— AID-Survey, Elgon-Nyanza, Vol. I & II, Marco Surveys. Nairobi 1962, Mimeo.

WILSON, F.: Kenya, A District Team at Work. Corona, Vol. 3, Aug./Sept. 1951.

WILSON, R. G.: Land Consolidation in the Fort Hall District of Kenya. J. Afr. Adm., Vol. VIII, No. 3, July 1956.